AF605790

Buddhist Cosmology

The Study of a Burmese Manuscript

Buddhist Cosmology

The Study of a Burmese Manuscript

JAMES EMANUEL BOGLE

Silkworm Books

ISBN: 978-616-215-122-4

First published in 2016 by Silkworm Books

104/5 M. 7, Chiang Mai–Hot Road, T. Suthep
Chiang Mai 50200 Thailand
P.O. Box 296, Phrasingh Post Office, Chiang Mai 50000
info@silkwormbooks.com
http://www.silkwormbooks.com

Design by Lisa Carta

Illustrations are either from the collection of the author or were specifically commissioned for the book. Some of the illustrations used were the author's art donated to the Walters Art Museum, Baltimore, Maryland, in 2010. Two exceptions are figures extracted and redrawn from *Trai Phum Book: Ayutthaya Manuscripts–Thonburi Manuscripts* (Amarin, 1999).

Photography by James E. Bogle and Preeda Phatharasathianchai initially, and others.

Front cover: Mount Meru cosmography from the manuscript.
Back cover: Maiden Fruit tree from the manuscript.

Title page: Birth Story – The Nemi Jataka. Drawing extracted from a Phra Bot, author's collection, by Khun Dennapa Sodsai, 2013.

This book is in memory of my parents,
James and Mary

my teacher,
Stanley Wollner

and my friend,
James Marsh Thomson

all of whom have departed and who now may know the truth.

CONTENTS

PROLOGUE

Buddhism 2
Theravāda Buddhism 3
Theravāda Buddhism in Asia 5
The Traibhūmikatha 6

PART 1 BUDDHIST COSMOLOGY

Overview 7
Are There Reasons for a Buddhist Cosmology? 8
The External Roots of Buddhist Cosmology 8
The Internal Roots of Buddhist Cosmology 10
A Cosmic Digression 13
The Jain Version of Jumbudvīpa 13
Introductory Thoughts 15
Rebirth Designations 16
The Three Worlds and the 31 Levels of Existence 16
Words 17
Yojana 17
Mahākappa and Celestial Years 18
Jhāna 18
The 31 Levels of Existence 18
Amplification of Level 31
Niraya: The Hells and Realm of the Damned 25
The Minor Hells of *Abhidharmakośabhāṣyam* 26
Geophysical Aspects of the Cakravāla 29
The Cakravāla 29
Plan Views 32
Mount Meru 33
Sattaparibhanda Golden Mountains and the Sīdantara Sea 35
Jumbudvīpa — Our World 39
Destruction and Recreation 40
Summation 40

PART 2 THE MANUSCRIPT: AN OVERVIEW

Origins 41
Burmese Manuscripts 42
Burmese-Thai Intersects 43
A Caveat 43
Iconography 43
Renderings 45
Notes on Translations and Script 45
Orientation 46
The Manuscript's Order of Presentation 47
Manuscript Pages in Order of Appearance 48

PART 3 MANUSCRIPT: THE WORLDS OF THE COSMOS

The Three Worlds and 31 Levels of Existence 57
Partial Translation of Manuscript Page 18: The Realms 58
The Heavens of the Cosmos 60

World One – 4 Heavenly Realms without Material Factors (*Arūpaloka*) 60
World Two – 16 Heavenly Realms with Material Factors (*Rūpaloka*) 62
The Five Pure Abodes (*Suddhāvāsa*) 62
Levels 10 to 20 – Heavens with Material Factors – 11 Realms 63
World Three – Which Includes 6 Heavenly Realms and 5 Other Realms (*Kāmaloka*) 64
The Lowest Five Levels of Existence 70
The Realm of the Asura (Level 28) (Manuscript pages 17 & 18) 70
The Realm of the Preta (Level 29) (Manuscript pages 21 & 22) 71
Niraya: The Hells of the Cosmos (Level 31) 73
Niraya: Translation of Manuscript Page 19 73
The Major Hot Hells 74
Other Hells 80
The Cakravāla's Cosmography: Translation of Manuscript Page 18 83
Mount Meru: The Great Cosmic Axle of the Cakravāla 86
Mount Meru 86
The Celestial Orbs 87
Sattaparibhanda: The Seven Mountain Ranges Surrounding Mount Meru 87
At the Base of Mount Meru 88
The Great Ocean and the Ānandā Fish 88
Jumbudvīpa's Cosmography: Translation of Manuscript Page 28 89
Jumbudvīpa: Plan and Major Elements 91
Major Elements 92
The Buddha with Devotees 93
Rivers and Lakes 94
The Himavanta Lakes 95
The Mountains Surrounding Anotatta Lake 96
Possible Geographic Aspects of Jumbudvīpa 98
Cakravāla 98
The Plan Of The Cakravāla 98
Trees and Other Features 100
The Cakravāla's Destruction 100
Translation of Manuscript Page 33 101
The Maiden Fruit Tree 103

THE ARHATS
Introduction 105
Arhats and Their Importance to Buddhism 107
The 10 Great Disciples of the Buddha 107
The 16 Arhats 108
The 93 Eminent Disciples of Buddha 109
Paintings of Arhats, Monks and Laity 111
Explanation of the Captions 112
Manuscript Page 35 113
Manuscript Page 36 114
Manuscript Page 37 115
Manuscript Page 38 116
Manuscript Page 39 117
Manuscript Page 40 118
Manuscript Page 41 119
Manuscript Page 42 120

Manuscript Page 43: Arhats and Male Religious Personages Accorded Pre-Eminence 121
Manuscript Page 44: Male Religious and Male Laity Personages Accorded Pre-Eminence 122
Manuscript Page 45: Notable Female Laity Personages 124

SELECTED HEAVENLY TEXTS
Caveats and Conditions 126
Six Texts Accompanying the Four Heavenly Realms without Material Factors (*Arūpaloka*) 130
Text Number 1 130
Text Number 2 131
Text Number 3 131
Text Number 4 131
Text Number 5 132
Text Number 6 132
Three Texts Accompanying the Heavens of the Five Pure Abodes 133
Text Number 7 133
Text Number 8 133
Text Number 9 134
A Text Accompanying Heavens 10 to 20 135
Text Number 10 135
Four Texts Accompanying Realms 21, 22, 23 and 25 136
Text Number 11 136
Text Number 12 136
Text Number 13 136
Text Number 14 137
Translator's Notes 137

EPILOGUE
A Polemic for the Theravāda Buddhist Cosmology 140

APPENDICES
Appendix A – Realms: Burmese Transliterations to Pali 141
Appendix B – Burmese Transliterations for Cosmological Names 142
Appendix C – Three Variations of the Rankings of 16–18 Arhats Charged with the Continuation and Protection of Buddha's Teachings 144
Appendix D – Alphabetical Listing of Monk/Arhat Personages with Rank of Importance to Buddhism 146

GLOSSARY AND ALTERNATE SPELLINGS 149

EXPLANATIONS OF SYMBOLS 152

ENDNOTES 153

BIBLIOGRAPHY 158

LIST OF FIGURES

PROLOGUE

Figure 1. Theravāda Buddhism in Southeast Asia and Sri Lanka

PART 1 BUDDHIST COSMOLOGY

Figure 1. Theravāda Buddhism in Southeast Asia and Sri Lanka 4
Figure 2. The Roots of the Theravāda Buddhist Cosmos 10
Figure 3. Plan View of Jain version of Jumbudvīpa 14
Figure 4. A Kṣuramārga Hell 26
Figure 5. Asipattravana Hell, a subordinate hell of Kṣuramārga Hell 27
Figure 6. Aya*ḥ*śalmalīvana Hell, a subordinate hell of Kṣuramārga Hell 27
Figure 8. The Eight Great Cold and Eight Hot Hells under Jumbudvīpa 28
Figure 7. A Hot Hell and its Usudas 28
Figure 9. The diameters of the Cakravāla and the Earth 29
Figure 10. A Sketch of the Cakravāla 30
Figure 11. A Section through the Cakravāla 31
Figure 12. An interpretation of the myriads of *Cakravālas* in the cosmic universe 31
Figure 13. Plan of the Cakravāla 32
Figure 14. Manuscript Plan of the Cakravāla 32
Figure 15. Configuration of Mount Meru according to the *Abhidharmakośabhāṣyam* 34
Figure 16. Configuration of Mount Meru according to King Ruang 34
Figure 17. Cross Section of the Sattaparibhanda Mountains 36
Figure 18. Manuscript Section of the Sattaparibhanda Mountains 36
Figure 19. Square configuration of the first four Sattaparibhanda Mountain Ranges 37
Figure 20. The Sattaparibhanda Mountain Ranges – annular configuration 37
Figure 21. India and Jumbudvīpa

PART 2 THE MANUSCRIPT: AN OVERVIEW

Figure 22. Sequential Arrangement of the Manuscript 47
Figure 23. Reproduction of the manuscript's drawing of Tāvatimsa Heaven 67
Figure 24. Painting of the inhabitants of Preta, Northeast Thai / Cambodian Manuscript painting 72
Figure 25. Phra Malai Visiting Preta, Thai Manuscript Painting 72
Figure 26. The Eight Major Hot Hells under the Continent of Jumbudvīpa 74
Figure 27. The Lokanta Hells (Manuscript page 32) 75
Figure 28. A Lokanta Hell from a Thai Manuscript 82
Figure 29. Mount Meru and its surroundings (Manuscript pages 16 & 17) 86
Figure 30. The Orbits of the Sun and Moon (Manuscript page 17) 87
Figure 31. The Black Chambers 88
Figure 32. The Great Ocean and the Ānandā Fish – A Vignette 88
Figure 33. Jumbudvīpa and Australia with their plans to scale 91
Figure 34. Jumbudvīpa's Plan (Manuscript page 29) 91
Figure 35. Specific Named Jumbudvīpa Locations 92
Figure 36. Major Elements of Jumbudvīpa's Plan 93
Figure 37. The Anotatta Lake and its rivers 94
Figure 38. The Lakes of the Himavanta (Manuscript page 30) 95
Figure 39. An Extrapolation of the Ganges River System 96
Figure 40. A Visualization of a possible geography of Jumbudvīpa 98
Figure 41. Manuscript Plan of the Cakravāla (Manuscript page 31) 99

Figure 42. Trees of the Cakravāla and the Lokanta hells 100
Figure 43. Maiden Fruit Tree (Manuscript page 34) 103
Figure 44. A Maiden Fruit Tree from a Thai Manuscript 104

LIST OF TABLES

PART 1. BUDDHIST COSMOLOGY
Table 1. The 31 Levels of Existence 19
Table 2. The Major Hot Hells 25
Table 3. The Four Continents, Their Inhabitants and Details 38

PART 3. MANUSCRIPT: THE WORLDS OF THE COSMOS
Table 4. The 33 Realms of the Cosmos 59
Table 5. Minor Hells; Their Names and Torments 80

PART 4. COSMOGRAPHY
Table 6. Destruction by Levels with the Various Forces of Destruction 102

PART 5. THE ARHATS
Table 7. The 10 Great Disciples of the Buddha 108
Table 8. A Ranking and Inclusion Listing of the 16 Arhats 109
Table 9. Eighty Male Arhat Personages 110
Table 10. Thirteen Eminent Female Personages 111
Table 11. Burmese Transliteration and Meanings 127

LIST OF TEXT BOXES

PROLOGUE

Buddha in Tāvatimsa Heaven 11
Birth Story (Jataka) 541 – The Nemi Jātaka – The lesson of Resolution 12
Mahāvīra 549–477 B.C.E. / Buddha 559–478 B.C.E. 13

PART 3. MANUSCRIPT: THE WORLDS OF THE COSMOS

Maitreya in Tusita Heaven 61
Phra Malai in Tāvatimsa Heaven 64
Garuda and Naga 65
Yama: The Ruler of Hell 71

PART 4. COSMOGRAPHY

An Offering Vessel 89
Rishis 100

PART 5. THE ARAHATS

Statues and Wax Images of Departed Monks 108

Acknowledgements

It goes without saying that an author not conversant with Burmese or Pali, or the other languages important to the Theravāda cosmos, and who is not a scholar of Buddhism owes much to others. The beginning of the work of this book began with the initial translation of parts of the manuscript by Ashin Sopaka, a Burmese Buddhist monk and scholar, on temporary assignment at Chiang Mai's Best Friend Library, a library focused on Burma. Akiko Thomson Guevara, my fictive niece, was instrumental in finding Ashin Sopaka. Through repeated urgings in person and by email Akiko urged me go to the library and seek translation help. I walked in the library's door and found Ashin Sopaka sitting serenely at a table, seemingly waiting for me to arrive.

Ashin Sopaka, a Burmese Buddhist monk and scholar with Khun Preeda Phatharasathianchai.

Khun Preeda Phatharasathianchai (Laa), in my employ, was a nearly perfect foil for my foibles and inadequacies. He greatly assisted me in Chiang Mai and Mae Sot, Thailand and Philadelphia, Pennsylvania with the photography, illustrations, typing, translation and computer work, and many other items in the development of this book. His knowledge of rudimentary written Burmese was helpful. His well-developed social skills were valuable to the realization of this book.

Two very knowledgeable gentlemen, who are also friends, contributed generously with their time and knowledge:

Dr. Leedom Lefferts assisted greatly in the clarification of this book's draft. His detailed critique of the draft was not only most helpful, but absolutely necessary for the realization of the book. He was also most generous and unstinting in sharing his knowledge and enthusiasm on the subject of Buddhism and Buddhist art.

Mr. Philip Blant, owner of 'The Centre of the Universe', Chiang Mai, accomplished the impossible in the identification and correction of my very great lacks of syntax, spelling, punctuations and other language foibles which have marked my literary passage for eight decades

The mistakes found, and there are bound to be some, are solely mine.

Of great help were the following:

Dennapa Sodsai, Art & Design, 40/1 Sridonchai Road, Chiang Mai developed the drawings for the Buddhist symbols and most of the numerous black and white drawings used throughout the book..

Sai Keing Hserng, a Burmese author, labored with the translations of the 'Cosmography' and 'Heavenly Texts', parts of the book. His products from those labors were most thoughtful and scholarly. His notes and insights improved the comprehension of the manuscript.

Thanks is also owed to:

The learned, acting Abbott, The Venerable Jarjoy Kitatipado, Wat Thaiwhatthanaran, Mae Sot, Thailand was instrumental in assisting with the understanding and translation of the arhat portion of the manuscript.

Dr. Bonnie Brereton, Chiang Mai, gave insightful advice and help at various points in the development of this work.

Dr. William Pruitt, Publications Administrator of the Pali Text Society, U.K. and Professor Sommai Premchit, Mahamakut Buddhist University, Chiang Mai, Thailand were instrumental in the conversion of Burmese nouns into Pali in Part 6 of the book.

Mr. Joel Akins, Editor, Silkworm Books, who put me back on the right course in a number of instances when I strayed.

Mr. John C. Mitchell, author and publisher, who suggested the book's title which, happily, replaced the convoluted working titles used previously.

Mr. David Lawitts who suggested and provided references.

Khun Theeravut and Khun Kowit of Sanguan Art, Nightbazaar, Changklan Road, Chiang Mai were very helpful in interpreting my sketches and developing them into illustrations that became works of art in their own right.

Rudolph Valentin Obmerga, my fictive grandson, proved most helpful with editing. He also possessed very useful ,and most helpful, computer skills which are apparently available only to the young who grew up with computers.are the states of minds and the rebirths of the inhabitants and deities.

Preface

A visit to the Asian Museum, San Francisco in 2004 began this effort. After looking at the manuscript I had brought with me, Dr. Forrest McGill, the Chief Curator, suggested that I should read the Reynolds & Reynolds book to understand the manuscript. This was the beginning of an arduous journey.

My wanderings in the Buddhist cosmos proved to be an awesome and time-consuming experience. The acquired antique Burmese cosmology manuscript was an invaluable tool to understand where to wander, and the limits of the paths.

This illustrated manuscript produced in the service of Buddhism is not a splendid artistic tool but a utilitarian one; designed to teach. This type of manuscript, a folding parabaik written on a mulberry paper, as studied in this book is no longer required for the transmission of Buddhism; its day has come and gone. The religion and literary traditions of Theravāda Buddhism, which spawned the manuscript, are very much alive.

In this effort, I realized that while the Buddhist cosmos has been described in the literature, it has not been the subject of much discussion. While there are great numbers of books on the Buddha, Buddhism and various aspects of Buddhist practice and thought, there are less than a handful of books in English which focus on the cosmos. In fact, of all the myriad of English language books which deal with Buddhism, only a few mention either the Hinayana or Mahayana Buddhist cosmos and even those do so only peripherally.

The manuscript, for reasons discussed in the book, has proven to be rare. The manuscript not only contains a colophon but also ventures into what appears to be rarefactive, and perhaps previously unknown, aspects of the Theravāda cosmos. These are the states of minds and the rebirths of the inhabitants and deities.

The efforts presented here are not a 'book' per se; in format it is more of a report than book: Let us begin.

Prologue

Before there were satellites, before there was Google Earth, before explorers ventured into the far reaches of the globe, men dreamed. These men's imaginations reached out past their limited worlds of distances hampered by a lack of swift vehicles, and terrain which was only accessible only by foot, horse, and small boats. These men dreamed of a perfect cosmos which answered their questions about their place in the schemes of things, much as you and I would dream of a perfect world, heaven, or even a terrible hell. We cannot fault them for their great, and much larger than life, thoughts. We have to admire them and we have to think about minds that would ponder such dreams and put them in writing.

The Theravāda Buddhist cosmos is not our world, though it bears some resemblance to our world, and enjoys some names of places known to us. This Theravāda cosmos is a world of wishes, dreams, and, most of all, cautionary tales of how Karma affects the ongoing lives of those who listen and read these stories. It is also a world of gods, deities, heavens and hells that are orderly, structured and segmented. It is a place of great and almost immeasurable periods of time. It is also the world of huge geological structures; oceans and mountains of immense dimensions and continents and other components extremely orderly in their placement and hierarchy. It is a world of great thought that beckons to be discussed, that needs to be examined and explored.

This is the world of the *Cakravāla,* the inhabited cosmologically unit we inhabit[1] and of *Saṃsāra,* which deals with the cycles of endless rounds of death and rebirth, in which humans exist. In this *Cakravāla* world, humans can transmigrate from one status to another, either better or worse, in the cycle of rebirth and death – depending on the results of their volitional actions, i.e. Karma. The reality of this perceived world universe is that it comes into being through Karma, and is destroyed by Karma, the actions and results of human activity.

This cosmos is a 'world system', one of many, nay, myriads of world systems. The singular one that is addressed is called 'Cakravāla'. All the cosmological world systems are subjected to cycles of development and decay. These cycles take eons and eons of time. Eventually, as decay deepens, lifespans shorten, morality disappears and destruction occurs by wind, water or fire. The world system dissolves and remains in a state of emptiness. Then, like a burned out forest, or the aftermath of a great volcanic eruption the emptiness is filled and life is recreated. There are parallels with today's knowledge that reflects this. Professor Akira Sadakata wrote:

> If we remove the graphic, the dogmatic, and the mythological from the expressions of Buddhist cosmologists, we are left with a series of concepts that resembles in no small way the conclusions of modern science.[2]

Buddhism

Buddhism, one of the oldest[I] of the world's great religions, is followed in age by Christianity and Islam. Buddhism is unique among these other religions in that it does not demand an exclusive allegiance, i.e. '*Thou shall have no other gods before me.*' Buddhism is basically a philosophy and makes no such demands. Buddhists do not believe as does the monotheist, that 'God' created the world and is relevant to salvation. Buddhism is not concerned with a God, or the world. Buddhism lacks the sacraments, rites, sacred institutions that can provide escape from continuing misery of *Saṃsāra*. There is no savior or helper; each one has to work out his own salvation and there is nothing outside the individual that can be of assistance.[3] Buddhism is concerned with man and how he lives his life, human suffering, ethics, morality, mediation and gnosis.[4]

Buddhism in its long history of two and one half millennia has realized two major philosophical forms.[II] These were the Hinayana and the Mahayana philosophies. Thomas Berry notes:

> Buddhism, as any vital spiritual or intellectual tradition, felt within itself tensions from a very early period. These tensions existed within an undivided community during the lifetime of Buddha and for some hundreds years after his death.[5]

At the Third Buddhist Council (ca. 250 B.C.E.[III]) the first major schism split the heretofore undivided community into two factions. This was followed by further splits — for example the Mahasanghika School fractured into six more schools between 280 and 240 B.C.E. Over time, other schisms and the schools of the Hinayana developed between the death of Buddha and the end of the first century B.C.E. These are thought to number between 18 and 20.

I "The orgins of Hinduism go back so far there are no written records of its beginings." It is believed to have orginated in the Indus Valley more than 3,000 years ago (Wilkinson 2010, p. 163).

II Vajrayana: Vehicle of the Thunderbolt, a form of tantric Buddhism that took hold in India as the Hinayana philosophy declined around the 6th century C. E., is sometimes considered a third philosophical form of Buddhism (see Lopez 2001, pp. 213–219). However, most writers believe it to be a form of Mahayana.

III B.C.E. = Before Common Era (Equivalent to B.C.) * C.E. = Common Era (Equivalent to A.D.)

The differences have been summarized in the terms 'The Great Vehicle' or 'Great Conveyance' for the Mahayana philosophical form and the 'Small Vehicle' or 'Small Conveyance' for the Hinayana form. The last appellation arrives from the adepts of Mahayana who believed that their ideals were superior to the older form which they derided. The use of Hinayana was meant to be derogatory. Mahayana needed to create an identity and make inroads into the followers of Theravāda and did so by declaring its superiority.[6]

The terms 'small vehicle' and 'small conveyance' are now commonplace and accepted; although Hinayanists prefer Theravāda. Even less pejorative are the appellations 'Southern Buddhism' for Theravāda practiced in southern and Southeast Asia and 'Northern Buddhism' for Mahayana as found in the northern Asian countries of Japan, Korea, China, Mongolia, Nepal, and Tibet. The differences appear vast, yet the relationship is close[IV] between these two philosophical forms.[7] The core difference was that the Mahayana followers believed that the Hinayana schools which focus on self-salvation was selfish instead of striving for the salvation of all beings. Over time the *Mahayana* Buddhist philosophy moved away from, and progressively deviated from the thoughts of Buddha, and split into many sects and even sub-sects.[8]

> Mahayana developed a vast body of metaphysical speculation and a huge pantheon. In place of the godless religion of the historical Buddha, the Mahayanists have myriads of godlike Buddhas in aeon of time.[9]

Theravāda Buddhism

Theravāda is the oldest and only surviving Hinayana school. Theravāda, a word from Sanskrit, means the 'teaching' or 'doctrine' of the elders. Theravāda was established in Sri Lanka in the 3rd Century B.C.E. by missionaries sent there by the Emperor Asoka of India who wanted to propagate Buddhism in neighboring countries.[10]

The twenty or so former Hinayana schools are thought to have represented the original and pure teachings of the Buddha. They were based on the sūtras which are based on the words spoken by the Buddha.

The Hinayana philosophy is one of self-liberation. All Hinayana schools had a realistic view of existence, with a goal of freedom and liberation from the constant recycles of death and rebirth. The attainment of Nibbana is the ultimate goal.

> The fundamental truths on which Buddhism is founded are not metaphysical or theological, but rather psychological. Basic is the doctrine of the "Four Noble Truths": (1) that all life is inevitably sorrowful; (2) that sorrow is due to craving; (3) that it can only be stopped by the stopping of craving; and (4) that this can only be done by a course of careful disciplined and moral conduct, culminating in the life of concentration and meditation led by the Buddhist monk.[11]

The ideal Hinayana personage is the monk who becomes an *arhat*, not the lay person. It is thought that the layman cannot attain Nibbana, which the arhat seeks to attain through his own efforts in renouncing the world.

IV Nine basis points unifying the Theravada and Mahayana philosophies were reached in 1967 at the First Congress of the World Buddhist Sangha Council.

Theravāda cosmology not only places man in the universe but also dictates his course of action through life. Forces from the cardinal points of the compass and from the planets and stars produce either prosperity or havoc. Thus, it is necessary for the state and individuals to be in harmony with these astral forces. There are lucky and unlucky days and many other rules which governed the individual and state – indeed, states were organized as small scale images of the universe; Angkor Wat in present-day Cambodia is one surviving example. The many Mount Merus built by Thai kings in the 18th and 19th centuries are others. Brahmin priests acted as intermediates between the state-individual, and forces of the universe as they do today.[V] The root of Theravāda cosmology is that of inevitable change via the consequences of Karma. Karma causes the creation and the dissolution of the world and of individuals.

To repeat: Theravāda Buddhism is the last remaining school of the Hinayana philosophy; the others have, over time, disappeared. It would seem that an anomaly exists, and either the word Hinayana or Theravāda should be discarded since the Hinayana form of Buddhism now contains only one school – Theravāda. However, as that may be, Lopez noted:

> As the last remaining school of the many Indian non-Mahāyāna schools, "Theravāda" is often mistaken as a synonym of "Hīnayāna."[12]

Figure 1. Theravāda Buddhism in Southeast Asia and Sri Lanka

V The Hindu Khmer kings used Brahmins as court advisors after the fall of Angkor Wat to the Thais in the 13th century the Thais kept the Brahmins so as not to cause unnecessary problems with the locals and then imported the Brahmins into usage at the Royal Thai court.

Theravāda Buddhism in Asia

Theravāda Buddhism links Sri Lanka in South Asia and the countries of Southeast Asia. Theravāda Buddhism has created a relatively homogenous religious culture that is universal to all countries that host the Theravāda religion – this despite their many ethnic, linguistic and secular cultural differences which exist in and between these countries.[13] In Southeast Asian countries Theravāda Buddhism is the main religion in Burma (Myanmar), Cambodia, Laos and Thailand. Other areas with pockets of Theravāda Buddhism are southern Vietnam where it is practiced by the Khmer Khom people, and the Chinese provinces adjacent to Burma and Laos where the Dai, Lieu, and Shan live. Theravāda Buddhism is also found in Bangladesh where groups of follower's live, and around the Chittagong Hill Tracts. Figure 1 shows the areas of the world where Theravāda Buddhism is currently practiced.

The roots of Theravāda Buddhism in Southeast Asia spring from Sri Lanka (Ceylon) which by happenstance after the demise of Buddhism in India[VI] became the keeper of Hinayana precepts. These are believed to be only a part of the total corpus of precepts and other doctrinal elements; they happen to be that part of the doctrine preserved in writing. The Buddhist precepts from Ceylon are written in the Pali language and consist of the Tripiṭaka (The Three Baskets) which address the three main aspects of Buddhist beliefs: the Vinaya, the rules for the Sangha and monastic order; the Sūtras, sermons for the laity; and the Abhidhamma, philosophical thoughts. Later, Thai King Ruang's 14th century Traiphum, dealing with the three worlds, was a major influence throughout Southeast Asian of Theravāda and its cosmology.

In Buddhist Southeast Asia the physical aspects of Theravāda Buddhism are manifest everywhere. Temples dot the landscape wherever self-sustaining cluster villages and cultivation are to be found. Rises in topography, small mountains and hills deemed to have merit, have stupas built on them which thrust into the sky and remind travelers of their faith.

Those engaged in Buddhist arts are also hard at work. Traditional Burmese, Cambodian, Lao and Thai painting, as with all Buddhist art, are primarily products of religious faith and not an aesthetic expression. Their purpose has always been to remind the faithful of the doctrines of the Buddha. The Thai author, Kamala Tiyavanich noted:

> In the nineteenth century there were monks all over Siam who were skilled artists and craftsmen. They passed on their knowledge to younger monks and novices. Buddhist artists did not sign their work, since they considered their efforts a religious effort. Personal fame was inconceivable to them.[14]

In Burma, Cambodia, Laos, and Thailand and the other places of Theravāda Buddhism, paintings and murals are among the many didactic efforts with the purpose of instructing viewers. To accomplish this, these art works had to be clear and their meanings unambiguous. Since viewers would probably be illiterate, they had to be familiar with the story before them, or be in the presence of monks or persons to explain the meaning to them. However, in the case of the manuscript which forms the foundation of this book, attempts to explain the cosmos, the segmented heavens and

---

VI Beginning in the 3rd century B.C.E. Theravāda existed in harmony for several centuries with Mahayana which was gaining more adherents displacing Theravāda and then itself being displaced by new forms of Hinduism (see Blanchard 1958, pp. 88–90.)

hells, the universe described and the significance of the illustrated personages, requires interpolation by a learned person to explain the meanings, and most of all the dreams, warnings, and admonitions the writing contained.

The Traibhūmikatha

'*Traibhūmikatha'* was written by King Lithai, the fifth king of the Ruang dynasty, the kings of this dynasty were rulers of Sukhothai, the capital of Thailand, some 700 years ago.[15] It later became known as the *Three Worlds According to King Ruang.* Ruang is a composite name given to members of the Sukhothai dynasty.[16] In the text that follows King Ruang and King Lithai have been used interchangeably; when referring to the author of the 'Traibhūmikatha'. The Thai text has been translated, first by Frank and Mani Reynolds (1982) and then by the Thai National Team for Anthology of ASEAN Literature (1985).

The Burmese manuscript discussed here has both writing and illustrations to convey its message. This manuscript dates from 1842, and it follows in abbreviated form much of King Lithai's Trai Phum. The message is divided into two sections; one section deals with the cosmos the other deals with personages. The cosmos part of the manuscript constitutes approximately 80 percent of the pages while the part that deals with personages consists of a sizable 20 percent. The manuscript does not deal with all the aspects of Theraṿāda cosmology; a number of items, mostly minor, have not been addressed.

PART 1

Buddhist Cosmology

Overview

Theravāda Buddhist cosmology, when compared with that of Christianity, is found to be segmented and complex. The Christian universe consists of one heaven, one hell and, in some cases, purgatory.[VII] Heaven and hell are one-way paths without return; only purgatory has the possibility of forgiveness and change.

By contrast, the Buddhist cosmos has multiple worlds (Cakravālas) each with heavens and hells which reflect the consequences of one's past deeds. Birth in the heavens, hells and other levels of existence, depends on the quality of one's previous lives which provide paths and directions to places of rebirth. The heavens and hells are transitory. In fact all the hells of the Buddhist cosmos are states of purgation. Over very long periods of time rebirth and change occur. These complex multi-tiered thoughts and places are reinforced by a physical cosmos of great dimensions that includes a complex geography of great oceans, rivers, forests, mountains, continents, suns and, moons.

King Ruang's *Three Worlds* described physical structures and distances in precise terms and measurements. Nothing in the heavens, hells, continents, and oceans was omitted; all are described in great detail. This part, however, is a fusion of a number

VII Dante's *Inferno*, a 14th century poem by the Italian Dante Alighieri, depicts nine circles of hell within the earth.

of serious sources meant to set the stage for understanding the Theravāda cosmos presented in the Burmese manuscript that is the subject of this book. Thus, many of the details that crowd the pages of *The Traibhumikatha* and its later iteration *Three Worlds According to King Ruang* are not included here.

The unreality, or hyperreality, of the cosmic structural universe that is provided by great levels of meticulous detail, might be felt to constitute a reason for dismissing the cosmology outright. However, if one looks at the underlying thoughts that stem from Buddhism; this "hyper-reality" becomes understandable as a need to proselytize the believer's acceptance and understanding of Karma and its transmigration, i.e. *Saṃsāra.*

Are There Reasons for a Buddhist Cosmology?

The basis of cosmology, for all philosophies and religions, is an attempt to answer age-old questions about man's place in life and the scheme of things. Buddhism does not hew to the belief of the existence of a creator god[VIII] which some view as a philosophy and some as a religion with all the tenets and practices of a religion laid down by its founder.[IX] Buddhist thought was formed during a period and in a place where Brahmanism prevailed. Brahmanism grew from Aryan beliefs and was based on the worship and rituals of the gods of creation and nature. Buddhist thought, with its roots in Brahmanism, began and established itself on a higher level with an overall philosophy which was intellectual and espoused a message that enjoyed great simplicity. Buddhism, in its purest form, in thought, meaning, and philosophical character, has no need for cosmology. However, in the early 5th, 4th and 3rd centuries B.C.E.[17] followers of Buddha, coming from a rich myth-laden environment of gods and deities did have a need – a very strong need – for such structural lore, which overwhelmed the pure philosophical character of Buddhism and led to a number of accommodations, one of which was a cosmology. This cosmology was also fed by the visionary experiences of the Buddha and his disciples.[18]

The External Roots of Buddhist Cosmology

The center of both the physical earth and the metaphysical spiritual cosmic beliefs that began and grew with the long history of the development of Hinduism also found a home in two heretical philosophies (to Hinduism) that developed in the fifth century B.C.E.: Buddhism and Jainism.

The Indian sub-continent historically has been characterized by an astonishing diversity of ethnic, geographical and religious influences. The tectonic structure of the Himalayan mountains, the great plains of the river Ganges, and the forests and seas, have had inputs into the development of Indian mythologies, and hence cosmologies. Beginning in prehistory Indian civilizations have given rise to a great mythology from which much of Asian cosmology has sprung. As has been nicely put:

> The mythologist studies myths as colorful pieces that form together

VIII "Whether or not Buddhism is also considered a religion is a matter of definitions and perhaps of no necessary importance" (de Bary 1972, p. xv).

IX "By contrast, (to philosophy) religion involves, *practice*, a way of living, or useful application in real life. The way a religion is practiced has to be based on a definitive canon or fundamental principle accepted as axiomatic, with a clearly stated goal" (Payutto 2007, p. 3).

> the beautiful mosaic of a people's cosmology, which is the totality of the world in which they live, their cosmos.[19]

In brief, India around 4000 B.C.E. was populated by an agricultural people, the Dravidians, who worshipped gods connected with fertility. Both phallic worship and the cult of the mother goddess are found in symbols and seals from that time. Plants were shown growing from female wombs, and bull's horns[20] were symbols of male virility. Human sacrifices were performed,[21] as in other cultures, to assure fertility. The center of Dravidian civilization was the Indus Valley where proto-cities were formed. A hundred years or so after the second millennium B.C.E. had passed, a wave of Aryans, a light skinned, hard drinking warrior people with mastery of horses and superior military skills, overwhelmed the agrarian dark skinned Dravidians and brought with them their own rough culture and gods. Unlike the Dravidians, the Aryans were a nomadic people and their gods were not concerned with fertility but the universal elements.[22] Among the Aryan gods were: Varuna, the prime mover of the universe; Prithivi and Dyas, the earth and the sky; Indra the storm god; Vayu, the god of air and wind; and other gods of elements.[23] Our knowledge of the Aryan gods comes from the Vedic hymns collected around 800 B.C.E. By the time of these hymns, Aryan culture had become more sophisticated and had acquired a moral preoccupation and a divine hierarchy. Toward the end of the Vedic period the gods morphed from simple personifications and responsibilities into more elaborate deities. The Brahmāic age had begun and the priests were hard at work assuming powerful roles in society. There was no single overriding deity and great questions arose such as whether Brahmā should be considered a new deity, or an old one.[24] After all, Brahmā was the first of the three great Hindu gods and was thought of as the Creator; the father of gods and men.[25] The debate continued, the priestly Brahmins grew stronger and, into this ecclesiastical stew of orthodoxy, came the clarity of the heterodox thoughts and teachings of the Buddha.

In the accommodations and blending of the Hindu Pantheon into the Buddhist cosmos the old deities and gods found new roles:

Brahmā, a Hindu god, and, as first of the gods, the creator of the universe[26] was accorded three of the middle-range cosmic heavens in the Theravāda cosmos for himself and his retinue.

Dharma Raja, personifying justice who appeared as a new god in Hindu times, became known by his ancient name, **Yama**, the god of death, presiding over the cosmic hells as the judge of the dead.[27]

Indra (Sakka), formerly the storm god and a Vedic deity, maintained that role in the Hindu pantheon as god of the firmament and a supreme deity.[28] In Theravāda Buddhist cosmology Indra is still a king, but now only of the lower Tāvatimsa Heaven and of 32 deities.

Māra, the god of death, may be the absorption of the Hindu god Mavi, the king of demons, into Buddhism; nevertheless he remains a deity and reigns over the highest heaven and the other realms of the sensuous sphere.[29]

As noted, Buddhism began, and remains, intrinsically of a philosophical character. To the laity of the time, this was a distant creed and difficult to comprehend without a god, or gods.[30] The laity was accustomed to gods-a lot of them-and so in time the age-old beliefs and gods crept in and became, in a way, homogenized into Buddhism with its new thoughts. The Hindu pantheon was not forgotten, but parts were incorporated to meet the needs of the popular new Buddhism. In a number of places Buddhism and Hinduism exist together in wonderful harmony. The Jatakas,[X] the stories of the Buddha's previous lives, melded well with Hindu beliefs of reincarnation.

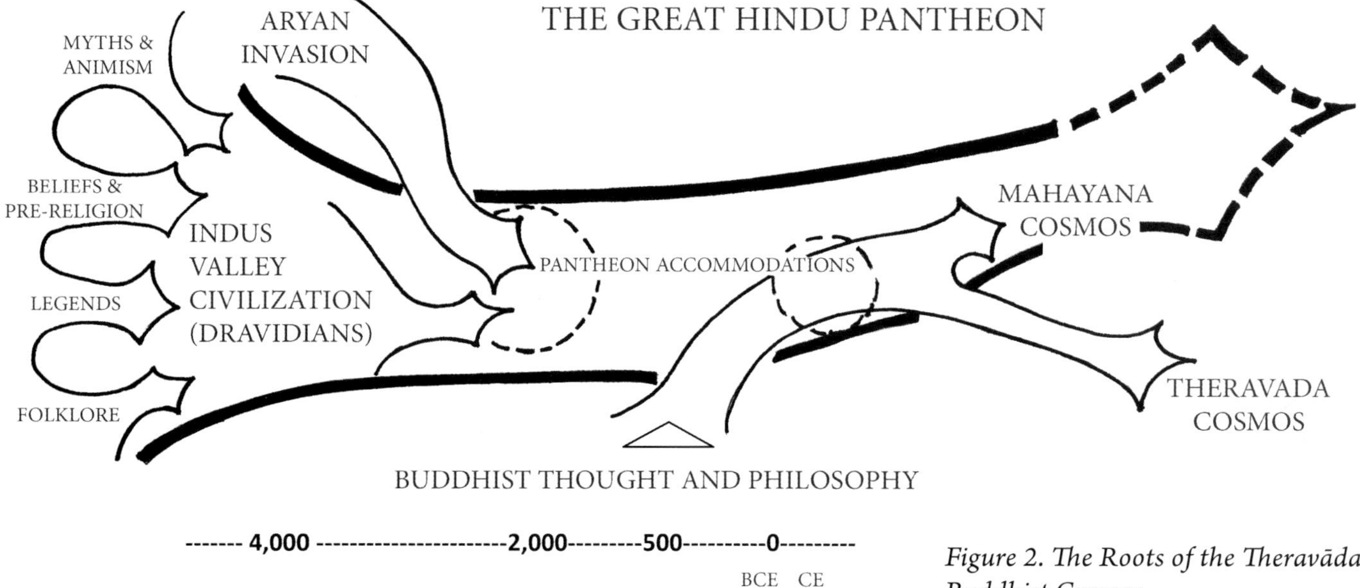

Figure 2. The Roots of the Theravāda Buddhist Cosmos

The Internal Roots of Buddhist Cosmology

Buddhist cosmology was also fed by the visionary experiences of Buddha and his disciples (Reynolds and Reynolds 1982, p. 15). Contributing to the complexity of Buddhist cosmology are what texts and tradition tell us about the meditative and visionary experience that was part of the Buddha's own road to Enlightenment.[31] His disciple, Moggallana, envisioned and visited the heavens which rewarded the just and the hells which punished sinners.[32] Stories, legends and new thoughts also emerged over the course of time which affected the cosmology. The Buddha espoused the meditation process and it was thought that, over time, meditation would lead to higher levels of heavenly rewards for holy ones who were disciplined enough to achieve it. Similar to changes in the growth of Jhānic[33] levels[XI] it could be reasoned that the number of heavens grew to cater to the growing levels of Jhānic meditative process which have

X The Jataka paintings portrayed scenes of religious history and are taken from a collection of 547 stories (some writers put the number at 550). These stories of the previous lives of the Lord Buddha are thought to be based on pre-Buddhist folk tales. The Jatakas which contained examples of moral approaches to life are a great source of material for teaching the precepts of Buddhism. Jatakas are recognized as part of Buddhist literature since they occur in the written canons of Buddhism. The Jatakas in reality are a reflection of transmigration of the ages past, as an authentic background to the historical life of Gotama.

XI In the early tradition of Theravāda Buddhism there were four Jhānic levels, eventually increasing to eight. (Reynolds & Reynolds 1982, p. 359).

affected and contributed to Theravāda cosmological development. The number of segmented hells may also have increased with time.

After the success of his Enlightenment, the Buddha worried that his Dhamma and message of self-abandonment were too difficult to explain. Then, as legend has it, the god Brahmā intervened. Pāli texts were used to introduce the existing gods into the narrative quite unselfconsciously and showed the gods to be part of their universe and legends. Thus the old gods Brahmā, Māra, and others contributed to the Buddha's story and forged a bond between the new philosophy of Buddhism and the older cults.[34]

In addition, there were tales of Moggallana, the Buddha, and Phra Malai visiting Tāvatimsa heaven and of King Nimi visiting the various hells. These legends reinforced the reality of the cosmos to the laity. (See the following text boxes: Buddha in Tāvatimsa Heaven[35] and the birth story (Jataka) of King Nemi's visits to the various hells.[36]

BUDDHA IN TĀVATIMSA HEAVEN

Buddha left the human plane after Enlightenment, to journey to Tāvatimsa Heaven, the domain of Indra and the abode his mother, Queen Maya, who had died in childbirth some 35 years previously. She now enjoyed the appellation of Maya Devi, signifying she was a goddess (Devi) Siddhartha. Her son, now the Buddha, came to teach her the wisdom of Abhidhamma. After a period of three months, residing and preaching in Tāvatimsa Heaven, he descended to the human plane.

Phra Vijaiyamuni of Wat Sri Mueng, Nong Khai, Thailand wrote:

On the full moon day of the eleventh lunar month, the Buddha closed the Lenten season with a ceremony and told Indra: "Today I shall descend into the world of humans."

Thereupon, Indra made three celestial ladders reaching to the earth. The golden one was on the right, the silver one on the left, and the jeweled one in the middle. Their lower portion rested at Sankassa, their upper portion rested on the mountain of King Simeru where Tusita heaven is located." The golden ladder was for the devatās, the silver one for Brahma's messengers, the middle one for the Buddha. When Buddha thus stood on the topmost rung on Simeru Mountain he could see the offerings of the devatās in different worlds and of human beings, and performed the yamakapātihāriya, or the miracle of the double appearances.

BIRTH STORY (JATAKA) 541 – THE NEMI JĀTAKA – THE LESSON OF RESOLUTION

On the appearance of his first gray hair, King Makhādeva of Videha, who ruled in the city of Mithhilā, passed the throne on to his son and became an ascetic. His son, grandsons, and all who followed and ruled, adopted this practice. All were born into Brahma's heaven until a king had a son named Nemi who in turn became king. From childhood Nemi was devoted to almsgiving and gave much away. In Tāvatimsa Heaven even the gods acknowledged Nemi's good works. Sakka (Indra), the ruler of this heaven, tried to explain to Nemi that the holy life was greater than almsgiving. Sakka ordered his charioteer, Mātali, to take Nemi and show him the various wonders and hells, including Vetarani, the river of hell. Mātali took his time to explain the sins that lead to those great hells. Finally Sakka sent a messenger and urged Nemi's return to heaven before he used up Nemi's life by showing him the endless wariety of the hells. The king returned to Sakka and was offered the uppermost heavens to reside in. However, with great resolution he declined, and returned to the world of men. He followed his ancestors' example; when the first gray hair appeared on his head he retired to a mango grove to develop the 'Four Excellences' which allowed him entry into Brahma's heaven. Nemi's son also followed his ancestors practice and became the last of the dynasty.

A Cosmic Digression

A Brahmāic-based cosmic commonality for Buddhism and Jainism sprang forth, almost in tandem, in the 5th and 4th centuries B.C.E. Jainism rejected the authority of the Vedas. The Jains also did not believe in a creator god and believed that divinity rests in every soul. Jains believed that liberation is obtained through right belief, right knowledge and right action. Mahāvīra, founder of Jainism, was a contemporary of Buddha.

There are a number of parallels between the founders of Buddhism and Jainism and also between the cosmoses of their religions.[37]

MAHĀVĪRA 549–477 B.C.E.
BUDDHA 559–478 B.C.E.

Both Mahāvīra and Buddha came from princely families
Both were born on the Indian sub-continent
Both renounced the world to seek release from material thoughts and possessions.
Both began their search for fulfillment as ascetic, mendicant beggars.
Both philosophies and religions that they espoused may be regarded
as heretical sects of Hinduism[38]

✲ ✲ ✲

The cosmologies that developed from their philosophies bear similarities.

	Buddhist Cosmology	Jain Cosmology
Worlds of Existence	3	3
Sacred Mountain	Meru	Meru
Height above Sea Level	80,000 yojanas	100,000 yojanas
Terraces	4	3
Principal Continent-Human	Jumbudvīpa	Jumbudvīpa
Heavens	26	16 or 26
Major Hells	8 hot, 8 cold	8 progressively colder

The Jain Version of Jumbudvīpa[39]

The disk of the Jain's Jumbudvīpa (Figure 3) may be compared with the Buddhist Jumbudvīpa presented later. The Jain Jumbudvīpa[XII] is bound within a blue border of water, with fishtails and water jars, *kalashes*, and a symbol of Brahmā at the four cardinal points. Within this border of water is set a yellow circle of temples and holy places. *Madhyaloka*, India, is represented by a trapezoidal land mass at the center bottom, emerging from the horizontal yellow band representing the Himalayas and bordered by two rivers flowing from the mountains, the Indus and the Ganges.

XII The Jain cosmos is made of three parts. The *madhyaloka,* or middle world, rests between the lower world, *adhholoka,* and the heavens, *urdhvaloka,* which comprise the Jain cosmos. The *madhyaloka* middle world is not only the smallest but also the most important, for this is where humans and animals live and where the tirthankaras, teachers who preach the dhammaa are born, and therefore, it is the only place where nibbana be obtained.

The adhaidvipa consisting of two-and-a-half continents of the *madhyaloka,* holds Jumbudvīpa. Jain illustrations show Jumbudvīpa as a cental disc surrounded by an ocean, then by a circle of land, another ocean and another circle of land and another ocean.

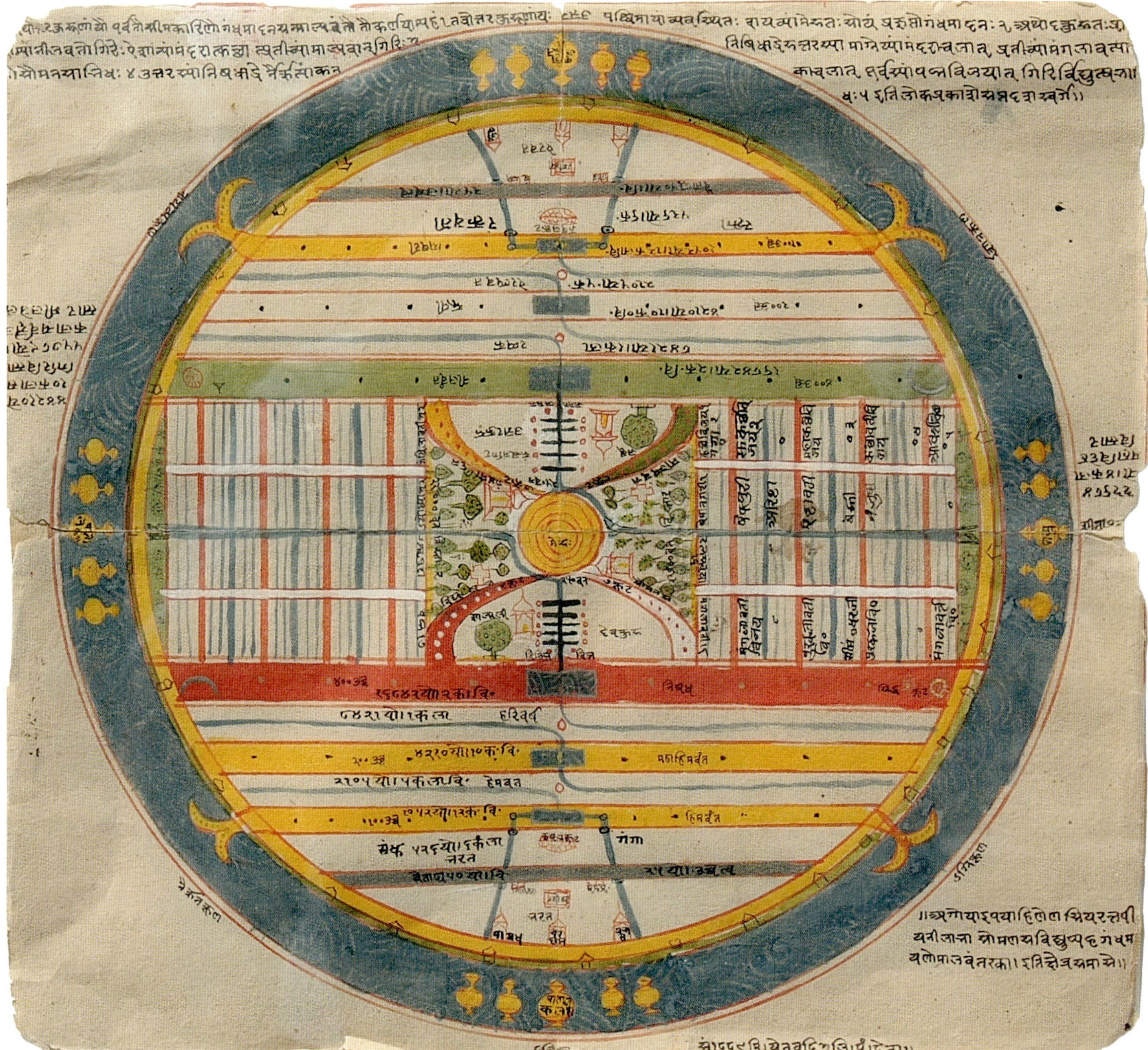

Figure 3. Plan View of Jain version of Jumbudvīpa

In the center Mount Meru is joined by two ranges of 'elephant tusks' mountains; the arcs of these tusks enclosed two Kurus (polities) of ancient India: One in the South where the Sālmalī tree is found and the other in the North where the Jambū tree grows. Forest areas lie to the east and west of Mount Meru. Six mountain ranges with parallel rivers run east to west cross the continent; three ranges in the north and three in the south. In the north the mountain range closest to Mount Meru is green; further north is an uncolored clear range and then a yellow range. In the south a red-colored range is found followed by two yellow-colored ranges to the south.[40]

Introductory Thoughts

The reality of the perceived Buddhist cosmos and universe,
comes into being through Karma —
and is destroyed by Karma — the actions,
and the results of, human activity.

The workings of Buddhist cosmology are vast, multi-faceted, interwoven and complex. The cosmology not only possesses psychological and spiritual worlds, but also physical ones, all of which are thought as being interlocked.[41] The following is brief, and only touches on aspects of the cosmos whose written development reaches far back into antiquity and was the work of many hands. The following recently penned caveats seem appropriate:

> When we read in the scriptures the complex numerical descriptions of the Buddhist universe, we feel overwhelmed. Even specialists find it difficult to understand the structure of Buddhist cosmology.[42]
>
> Disagreement concerning the nature and extent of the universe constitute a focus of theological debates which permeates Buddhism at every level.[43]

To know the cosmos, the following points (uncommon to Western and other religions) are helpful. Buddhist thought sees time as cyclical rather than linear. Hence:

- All the heavenly deities, or gods, live very long lives, but are not immortal and are subject to rebirth.
- The inhabitants of the hells are in a transitory status; albeit for extremely long periods of time; however, their time in hell is not permanent.
- The only permanent cessation of the cycle of life, death and rebirth (*Saṃsāra)* comes upon attaining Nibbana.
- The cosmic universe is subject to the same cyclical events as other aspects of Theravāda Buddhism and, after eons of existence, will end when the eon of dissolution occurs.

Buddhist cosmology has as its base a world system, an inhabited universe called *Cakravāla,* which provides, and accounts for, karma and the structure of rebirths. The Buddhist cosmos has both abstract and concrete components. In Buddhist cosmology myriads of other similar systems exist throughout the cosmos, but only one main path of ethereal thought applies to all these systems. This is the *Cakravāla* – the Buddhist universe which deals with the cycles of endless rounds of death and rebirth, *Saṃsāra,* in which humans and others exists.

The *Cakravāla* itself passes through four ages; creation, habituation, cataclysmic events and nothingness.[44] After eons of life the gods are subject to the same cycle of death and rebirth. The luster fades, the thrones become uncomfortable, bodies perspire and age, servants become indifferent and the palaces disintegrate.vision of the next lifetime in a lower realm occurs and with it the realization that one's past life whose time was squandered in pleasure was the cause.[45]

Rebirth Designations

Humans can transmigrate from one status to another, either for better or for worse, in the cycle of rebirth and death – depending on the results of their volitional actions, i.e. karma. Karma can lead to six destinations in which, depending on one's karma, it is possible to be born as:

- Devas (Gods – major and minor)
- Humans
- Animals (all kinds)
- Asura (Titans)
- Hungry Ghosts (Pretas)
- Inhabitants of the hells

The Three Worlds and the 31 Levels of Existence

The following discussion comes mainly from King Ruang's 14th-century sermon. Reynolds and Reynolds, in their translation stated:

> The Trai Phum Phra Ruang is a sermon that culminates a long history of visionary and cosmological literature within the Theravāda Buddhist tradition. It presents, in an especially vivid and concise form, the religious universe within which Thai Buddhists have traditionally lived.[46]

The world system of the cosmos has three worlds or realms.[47] Within these three worlds Theravāda cosmology indicates there are 31 levels of existence which include 26 heavens, and levels for humans, animals, ghosts, hells and others. The following tabulation places these levels of existence into the three worlds or realms:

The World of Nothingness, or No Form (Arupadhātu)

4 Heavens

The World of Form (Rūpadhātu)

16 Heavens

The World of Desire (Kāmadhātu)

6 Heavens

1 of Human Beings

1 of Asura

1 of Hungry Ghosts (Preta)

1 of Animals

1 of the Hells

The heavens are numerous, segmented, and transitory places, and in a number of cases overlap with only slight distinctions in nomenclature and reason for the separations. The heavens are also separated by great distances. Some of these distances are mind boggling as are the sizes given for their inhabitants. Beneath the 26 heavenly realms there are five realms, each of which is concerned with a singular class of inhabitants.

The Theravāda cosmos has eight major hells and numerous subsidiary ones, the major hells and all of the minor ones are all confined to a single realm (*Niraya)* and are also segmented and transitory. However, much more attention is given to this single realm than all others. This is a reflection of, and teaching tool for, the results of karma on life in the hereafter.

Words

Three terms which need to be understood to know the cosmos are: yojana, which deals with measurement; Mahākappa and Celestial Years, which deal with time, and the last, Jhāna, which deals with meditation.

Yojana

The *Abhidharmakośabhāṣyam*[48] and various versions of the Traiphum[49] give a description of the universe. The unit of measurement is the *yojana,* an ancient Indian term of measurement, which equates to the distance a cow yoked to a cart could pull, or the distance an army could march in one day, or a relationship to the height of a man. There are a number of estimates on just how far a *yojana* would be, ranging from 4 to 20 kilometers.[50] In truth no one knows the exact distance.[51] The great dimensions of the elements of the cosmic universe are overwhelming. Therefore, the distance used in this book for a yojana is the smallest one of four kilometers. One yojana has 8,000 *wa.*[52] The distances of the heavens from the Cakravāla, shown in Table 1, are extraordinary.

As an example, the distance from the earth to the sun averages 150 million kilometers,[53] or 37,500,000 yojanas. This distance is called Astronomical Unit. Eleven of the Theravāda Buddhist heavens when measured from the surface of the cosmic ocean, exceed that figure. For instance, the Heaven of the Highest/Supreme Devas (gods), *Akaniṭṭha* (Level 5), is over 8,948 astronomical units' distance from the ocean surface of the *Cakravāla.* See the following sketch.

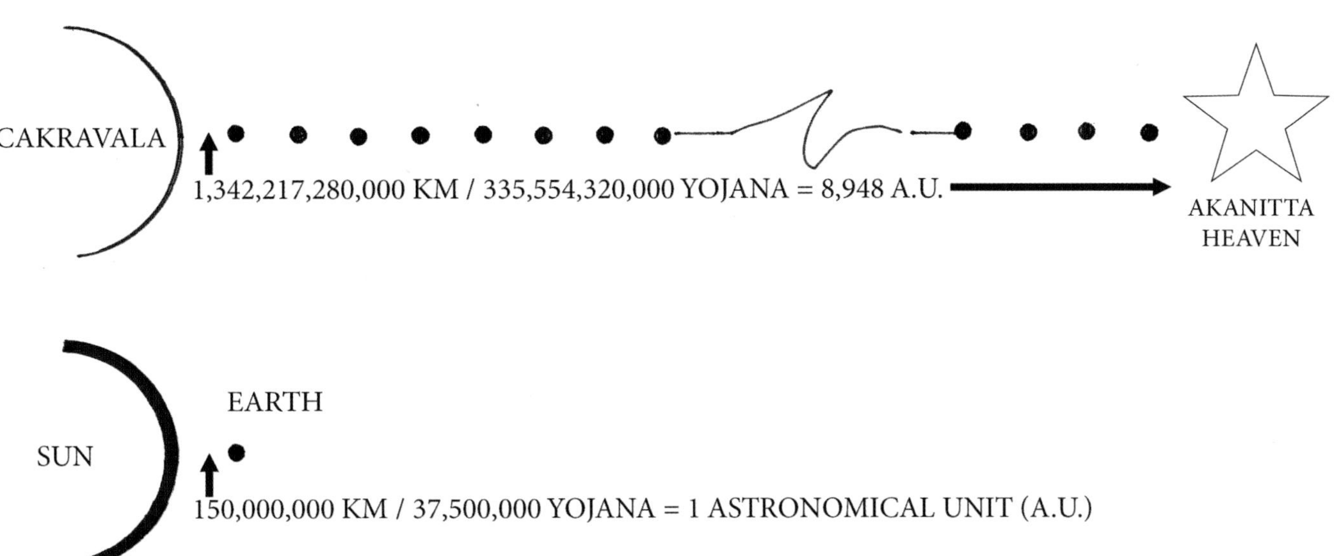

Mahākappa[54] *and Celestial Years*[55]

Mahākappa, Mahākalpa or Kalapa, is an ancient Indian measurement used to calculate great and almost immeasurable periods of time. One Mahākappa has been described as the duration of a day for the deity Brahmā, which is equal to 4,320 million human years.

The Buddha used the parable of a hill for comparison:

> Suppose there was a solid mass, a cube of rock one yojana per side (four kilometers) and every hundred years a man was to stroke it once with a piece of silk. Then that mass of rock would be worn away sooner than an aeon would be past.

The most common values[56] given for Mahākappa, are10^{51}, 10^{59}, or10^{63}

Celestial Years have been defined as 50 earth years equaling one celestial day, or to put it another way, 18,250 earth years equal one Celestial Year.

Jhāna

The Jhānas are defined by the Buddha as Right Concentration. Basically, Jhāna is a meditative state of profound quiet and utter concentration in which the mind becomes fully absorbed in the chosen object of attention and is characterized by non-dual consciousness.[57] It requires the ability to concentrate for long periods of time without becoming distracted. In the early tradition of Theravāda Buddhism there were four Jhānic levels, each of increasing ability to concentrate, these four levels eventually increased to eight.[58] The Jhānas are described many times in the Pāli Canon and they are 'elucidated' in quantities of post canonical Theravada literature.[59] In the Theravāda cosmos Jhānas have transcendent core roles. Levels of successful Jhāna meditation combined with cognition are the thresholds for entry after death, to the various higher heavens where the wait for rebirth begins. The Jhāna state itself cannot lead to enlightenment, as it only suppresses defilement. It does cultivate wisdom which acts on closure of defilement leading to Nibbana.[60] With the development of Jhānic thought came the possibility of both death and rebirth in the various realms of the higher Jhānic heavens.[61]

The 31 Levels of Existence

The 31 planes or levels of existence are shown in Table 1.[62] They begin with the highest level and descend to the lowest. This sequence is common to both Burmese and Thai manuscripts[63] and conforms to the arrangement of this Burmese manuscript, wherein the heavens are the overture and all else follows. This arrangement is the reverse of King's Lithai's Traibhumikatha,[64] which begins with the hells and moves upward to the heavens. The lifespan, condition of inhabitants, Jhānic level (where relevant), and Pali names are given in the Table. The Table also contains notes of interest about the various levels of existence. The sources used for this Table are not always in agreement and accommodations have been made in its development. Please keep in mind that Table 1 is a synopsis of lengthy data from various sources.

TABLE 1. THE 31 LEVELS OF EXISTENCE

WORLD OF NON-FORM *ARŪPALOKA*

Access to the four highest levels is gained by great mental attainment and mastery of the higher levels of the Jhānas at the time of death. Time spent in these worlds before rebirth is almost immeasurable and verges on infinity. A body is not present; only the mind is viable, which is unable to hear the Dhamma.

Level	Condition of Inhabitants	Mahākappa	Jhānic Entry Level	Distance Above Great Ocean	Height of Inhabitants
1	**Neither-perception-nor non perception** ***Nevasaññā-nāsaññāyatana*** These formless beings are at the threshold of physiological response; they cannot realize 'perception' but are not totally unconscious	80,000 (The Mahākappa of this heaven has also been placed at 84,000)	8th		
2	**Nothingness** ***Ākiñcaññāyatana*** These formless beings contemplate the thought "that there is no such thing' — a form of very subtle perception	60,000	7th	There is no physical site factor as these four realms have no location	There are no bodies, only minds
3	**Infinite consciousness** ***Viññānaññcāyatana*** These formless beings meditate on their consciousness as infinitely pervasive	40,000	6th		
4	**Infinite Space** ***Ākāsānaññcāyatana*** These formless beings meditate upon space as infinitely pervasive	20,000	5th		

WORLD OF FORM *RŪPALOKA*

The first of the physical realms where location and bodies exist. While these bodies have form they are not visible to the inhabitants of the next lower realm (Kāmaloka). There are no sexual distinctions, or desires for pleasing the senses; their minds correspond to the Jhānas.

Level	Condition of Inhabitants	Mahākappa	Jhānic Entry Level	Distance Above Great Ocean	Height of Inhabitants
5	**Highest/Supreme Devas (gods)** ***Akaniṭṭha*** The highest of the Rūpadhātu heavens where all devas enjoy being of equal rank. When Sakka (Indra), lord of Tāvatimsa heaven, dies he will be reborn here.	16,000	5th	335,544,320,000 yojana	16,000 yojana
6	**Clear Sighted Devas** ***Sudassī*** These clear seeing inhabitants live in a heaven similar to that of the Akaniṭṭha heaven (above).	8,000	5th	167,772,160,000 yojana	8,000 yojana
7	**Beautiful Devas** ***Sudassā*** The heaven of beautiful devas is reserved for anāgāmins, those who have eliminated the first five fetters of life; they will be reborn in one of the five 'pure abodes' and become an *arhat*.	4,000	4th	83,886,080,000 yojana	4,000 yojana
8	**Serene devas without troubles** ***Atappā*** This is the heaven of untroubled devas whose company those in the lower heavens wish for and envy.	2,000	4th	41,943,040,000 yojana	2,000 yojana
9	**Devas not falling from prosperity** ***Avihā*** This heaven is the most common place of rebirth for *anāgāmins*. Many achieve arhatship with a high Jhāna in this heaven, but even then some of those who pass are reborn in the higher heavens (*Sudassi, Sudassa, Atappa*) until they at last achieve rebirth in *Akaniṭṭa* heaven or *Nibbana*.	1,000		20,971,520,000 yojana	1,000 yojana
10	**Devas without perception** ***Asaññasatta*** Not unlike devas in the 'formless realms' found above, devas in this realm, not wanting to have the dangers of perception, have achieved states of non-perception in which they exist for a time. Eventually, they re-enter a state of perception and fall into a lower state.	500	4th	10,485,760,000 yojana	500 yojana
11	**Devas who are fruitful** ***Vehapphala*** The realm of devas who have 'great fruit'.	500	4th	5,242,880,000 yojana	250 yojana

12	**Devas of unlimited splendor** ***Subhakiṇṇā*** This is the heaven of total splendor and beauty.	64	3rd	655,360,000 yojana	125 yojana
13	**Devas of unlimited beauty** ***Appamāṇasubhā*** The heaven of limitless beauty where the inhabitants possess the virtues of great wisdom and learning.	32	3rd	327,680,000 yojana	125 yojana
14	**Devas of limited beauty** ***Parittasubhā*** This is the heaven of inhabitants of limited beauty.	16	3rd	163,840,000 yojana	125 yojana
15	**Radiant Devas** ***Ābhassarā*** This is the heaven of devas with splendor.	8	2nd	81,920,000 yojana	125 yojana
16	**Devas of unlimited splendor** ***Appamāṇabhā*** The inhabitants in this heaven possess limitless light and splendor, something on which they meditate.	4	2nd	40,960,000 yojana	125 yojana
17	**Devas of limited splendor** ***Parittābhā*** This is the heaven of devas of limited light or splendor.	2	2nd	20,480,000 yojana	125 yojana
18	**Mahabrahmā – the great Brahmā** ***Mahābrahmā*** The heaven of the 'Great Brahmā', once believed to be the creator of the world, who enjoyed many grand titles. He fell into this lower status through the exhaustion of his merit. He was reborn in a lower heaven and, forgetting his past, now believes that he has come into existence without cause. He has no knowledge of the upper heavens.	1	1st	10,240,000 yojana	1.5 yojana
19	**Deva priests & ministers of Mahabrahmā** ***Brahmaparohita*** The ministers and priests of Brahmā, who once belonged to the higher heavens, were born as companions to Brahmā in this lower heaven. Since they are a reflection of his thoughts and desires for companions, Brahmā regards himself as their creator. The ministers and priests also regard Brahmā as their creator. Since they have been reborn in this lower heaven they have some memory and teach the doctrine of Brahmā as a revealed truth.	0.5	1st	5,120,000 yojana	1 yojana
20	**Devas in the retinue of the Mahabrahmā** ***Brahmapārisajja*** The deities in this heaven belong to the assembly of Brahmā. They act to support Brahmā and the doctrine.	0.3	1st	2,560,000 yojana	0.5 yojana

WORLD OF DESIRE *KĀMALOKA*

Those born in this realm enjoy different degrees of happiness, but all are under the hegemony of Māra, which causes desire itself, the cause of suffering.

Level	Condition of Inhabitants	Mahākappa	Jhānic Entry Level	Distance Above Great Ocean	Height of Inhabitants
21	**Devas who find pleasure made by others** ***Paranimmita-vasavattī*** Māra is the lord of this heaven. Māra personifies delusions and attempts to keep all beings in the grip of sensuous pleasures and passions that overwhelm and hinder good from coming forth. Māra attacks the Buddha while he mediates in his quest for enlightenment. Part of these temptations included having his daughters tempt the Buddha as he mediated under the Bodhi tree.	16,000 *292,000,000	None	1,280,000 yojana	8 yojana
22	**Devas who enjoy their own works** ***Nimmānarati*** The devas of this heaven are capable of making an appearance that pleases themselves.	8,000 *146,000,000	None	640,000 yojana	4 yojana
23	**The birthplace of Bodhisattvas** ***Tusita*** Thought of as the most beautiful of the celestial worlds, Devas and other Bodhisattvas are born here before their next life as a Buddha. The Tusita palaces are made of gold and silver, with enclosing walls of crystal. The deva who rules is Santusitadevaraja. Maitreya, a bodhisattva who resides here, awaits rebirth as the new Buddha to renew the teachings.	4,000 *73,000,000	None	320,000 yojana	2 yojana
24	**The level of Yāmā Devas** ***Yāmā*** This is the heaven without fighting since it is disconnected from the great massif of Mount Meru. The Yāmā Devas live in the air and ares free of trouble. Palaces are of gold and silver surrounded by crystal walls. The king is Suyama-devaraja.	2,000 *36,500,000	None	160,000 yojana	1 yojana

25	**The home of Indra and 32 Devas** ***Tāvatimsa*** Tāvatimsa heaven, one of the lower heavens, is located on the top of Mount Meru. This heaven is thought to be one of the loveliest; it is the home of Indra (Sakka) and 32 Devas. It is also the site of the Culamani Chedi which houses the hair (or tooth) relic of the Buddha. This heaven was visited by the Buddha who spent time with his mother. Phra Malai also visited this heaven and here met Indra and Maitreya; the Buddha of the future. (The description of Tāvatimsa Heaven in most manuscript accounts is the most voluminous and detailed of all the heavens).	1,000 *18,250,000	None	On the summit of Mount Meru 80,000 yojana above the ocean	0.75 yojana
26	**The four kings of the cardinal quarters** ***Cātummahārājika*** The four kings who rule and stand guard over the four quarters of the universe, each resident in his quarter exist at this level. They live in palaces on the slopes of Mount Meru. The kings also reign over the Yakshas – ogre-like beings and sometime referred to as giants.	500 *9,125,000 	None	On the terraces of Mount Meru	0.75 yojana for Kings 0.5 Yojana others
	ABOVE ARE THE LEVELS OF THE HEAVENS (26) AND THE HOMES OF DEVAS – ALL MOST DESIRABLE FOR REBIRTH				
27	**The level of human beings** ***Manussa*** The details of the various abodes of humans vary as does the descriptions of the inhabitants (see Table 2).	Indefinite			

BELOW THE HUMAN LEVEL ARE THE LEVELS OF GREAT STRESS: THE LEVELS OF THE ASURA, PRETA, ANIMALS AND HELLS – THE MOST UNDESIRABLE FOR REBIRTH.

28	**The level of fallen devas and demons** ***Asura*** The Asura, demigods, were deposed from *Tāvatimsa* Heaven for actions based on jealousy and drunkenness. They are always fighting to regain their lost kingdom on the top of Mount Meru but are stopped by the four kings of the cardinal quarters who reside in palaces on the terraces of Mount Meru.	Indefinite
29	**The level of the hungry ghosts** ***Preta*** The inhabitants of this level are born here after a life of wicked deeds, primarily great envy of others and dismissal of the dharma and the four noble truths. As the result of their bad Karma, they have a great craving for particular unattainable foods or objects. 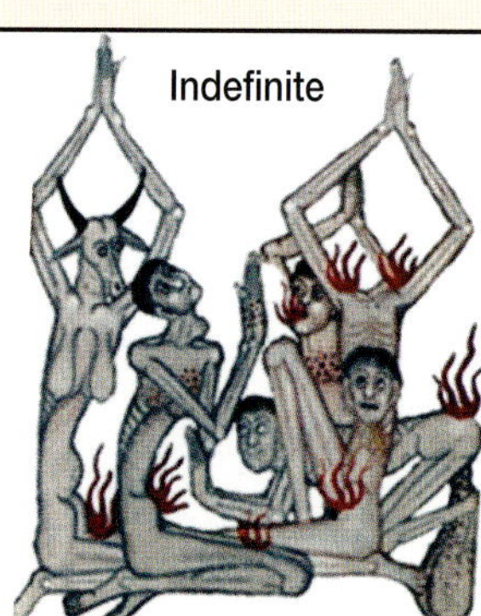	Indefinite
30	**The level for all animals** ***Tiracchāna***	Indefinite
31	**The hell realm of the damned** ***Niraya*** The hell world of the damned is the lowest level of existence. Hell is one of the more detailed and segmented realms; much more is known (and discussed) about the trials and tribulations of these inhabitants, than those in the all the other 30 levels of existence.	Indefinite

Amplification of Level 31– Niraya: *The Hells and Realm of the Damned*

Theravada Buddhism as expressed in the work of King Lithai suggests there are eight hot hells these are as follows:

TABLE 2. THE MAJOR HOT HELLS

No.	Name	Alternative Spellings	Type of Hell	Alternative Description
1.	Samjiva	Sañjīva	Death-Life Cycle	Constant repetition or Reviving Hell
2.	Kalasutra	Kāsutta	Black wire	Black Thread Marks
3.	Samghata	Saṅghāta	Stone slabs	The Crushing Hell
4.	Raurava	Rorva, Roruva	Groans and Moans	Lamentation or Screaming Hell
5.	Maha Raurava	Mahāraurava, Mahāroruva	Great Groans and Moans	Great lamentation or Great Screaming Hell
6.	Tapa	Tapana	Fiercely burning fire	Scorching heat
7.	Maha Tapa	Pratāpana, Mahātapana	Great Fiercely Burning Fire	Fiercely scorching heat
8.	Avici		Hell without interruption	The Uninterrupted Hell

The suffering of the inhabitants in the hot hells does not exhaust the myriad of possibilities of consignment to torture.

Some sources described

> five hundred or even hundreds of thousands of different Narakas [hells] . . . The sufferings of the dwellers in Naraka often resemble those of the Pretas and the two types are easily confused. The simplest distinction is that beings in Naraka are confined to their subterranean world, while the Pretas are free to move around.[67]

In addition, each hot hell has 16 minor hells (*Usuda*), four on each side which differ in use for each major hell. In one version, *Samira*, the first major hot hell, has the following 16 Usudas and specific reasons for incarnation.

The descriptions of what occurs in the *Abhidharmakośabhāṣyam* hells are most disturbing and awesome. These minor 5th century hells enjoyed currency in the 19th century. The Kṣuramārga Hell, with the great road of razor blades, has subordinate hells which are illustrated in the Thai Phra Malai manuscript paintings below. (These figures show the interpretation of hell as it was probably thought of when the Phra Malai mid-19th century and early 20th century manuscripts were painted.) The scenes of hell are most graphic, and are meant to be so. There are vultures feasting, arms and legs hacked off, and other methods that would induce extreme pain. These Thai Phra Malai manuscripts were commissioned by patrons and given to the temples to make merit for their donors. The treatment of the wrong doers varies with the artist and the depictions can be very imaginative.

The Minor Hells of Abhidharmakośabhāṣyam

The *Abhidharmakośabhāṣyam of Vasubandhu* identifies only four types of minor (*usuda*) hells with Kṣuramārga hell having two subordinate hells. Reasons for incarceration in these hells is not provided.

- *Kukūla*
- *Nadī Vaitaraṇī*
- Kunapa
- *Kṣuramārga*
  - Asipattravana
  - Ayaḥśalmalīvana

Figure 4. A Kṣuramārga Hell[68]

This Thai manuscript painting shows the Kṣuramārga Hell in which the sufferers loose their flesh. Birds with iron beaks tear out the eyes and flesh of the sufferers and eat them.

Figure 5. Asipattravana Hell, a subordinate hell of Kṣuramārga Hell[69]

The Asipattravana Hell is a forest with leaves like swords which cut the body for the Śyāmamaśabala dogs to eat. This paintings from a Thai manuscript shows one of the Yama servants cutting a sinner's body and the small, active dogs eating human limbs.

Figure 6. Ayaḥśalmalīvana Hell, a subordinate hell of Kṣuramārga Hell[70]

The Ayaḥśalmalīvana hell is a forest of trees with long thorns which turn down when climbing up, and turn up when climbing down.

This Thai manuscript painting shows the karmic consequences meted out to those who are governed by sexual passion, especially adultery. A woman at the top of a thorn tree is grasping at a suitor who tries to climb to her. In the process the suitor has his flesh lacerated by the thorns, while also being attacked by dogs and the Yama warden. On reaching the top he finds his that his lover is on the ground and he has to climb down only to find her back up in the top of the tree; this revolving, never successful pursuit endures for millions of human years.

Snelling noted:

> The human imagination seems to become unusually inspired when it comes to devising extreme forms of horror and torment-and the early Buddhists were no exception.[71]

Figure 7 shows one interpretation of a Hot Hell and the minor hells (all are drawn without roofs). The description of the major and minor hells given in the Traibhumikatha is quite precise. The major hells have hot iron bottoms, are 100 yojana on a side, walls, floors and ceilings are nine yojana thick. The major hot hells are unfit for the guardians to live; they reside in the minor hells. The hells are filled with sinners who provide the fuel for the flames.[72]

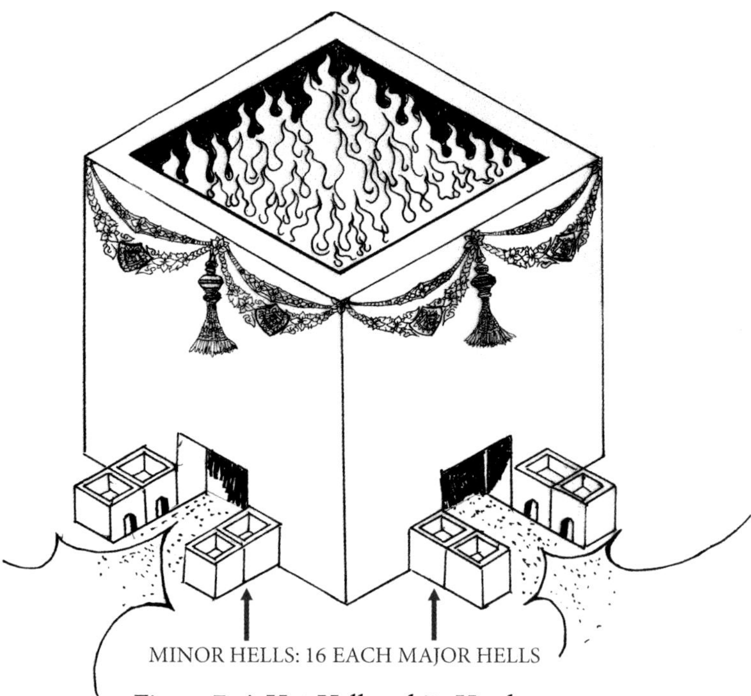

Figure 7. A Hot Hell and its Usudas

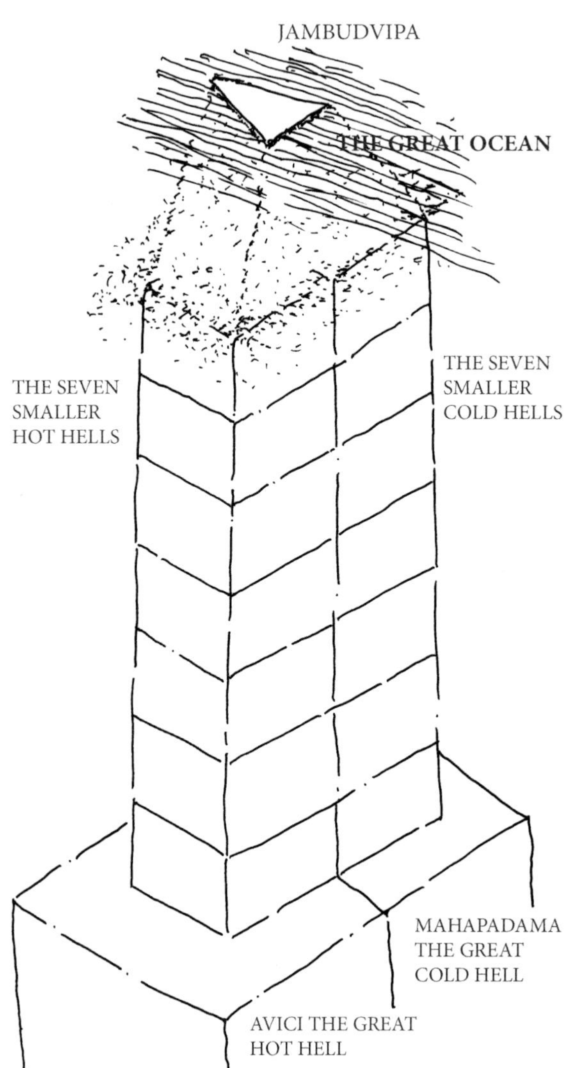

Figure 8. The Eight Great Cold and Eight Hot Hells under Jumbudvīpa

It is worth noting that the much older writings the *Abhidharmakośabhāṣyam of Vasubandhu* indicates that there are eight major hot and cold hells, and a number of minor ones. The most accepted version of the hot and cold hells is that the hells are layered upon each other under the continent of Jumbudvīpa. Both the Avici hot hell and Mahāpadama cold hell are of immense size. Jumbudvīpa is smaller than any of the hells and the difference is explained by the *Abhidharmakośabhāṣyam*: 'The continents, like piles of grain, are wider at their bases.'[73]

Cold Hells

1. Arbuda – The hell of blisters.
2. Nirarbuda – The hell of Burst Blister.
3. Aṭaṭa – The hell of shivering.
4. Hahava – The hell of Lamentation.
5. Huhuva – The hell of Chattering Teeth.
6. Utpala – The hell of the Blue Lotus.
7. Padma – The hell of the Lotus.
8. Mahāpadama – The hell of the Great Lotus.

Geophysical Aspects of the Cakravāla

The structural and physical aspect of the cosmos has been described, and elucidated in great detail. In 1871 one Victorian author, fluent in Thai,[XIII] noted:

> In the "Traiphoom" the system is elaborated in a most tedious manner and the strictest measurements are given of everything and place referred to.[74]

Nothing of importance is left to vagueness. Most unnecessary items are well described and measured. In dealing with the cosmologies that stem from the ancient Vedic writings, one is confronted with great dimensions, that considering today's realities, can only be viewed as unreal. Indeed, Professor Sadakata noted in the foreword of his book on *'Buddhist Cosmology'* that, for today's reader

> Buddhist teachings about the universe's structure may appear to be absurd.[75]

My intent in describing the physical world of the cosmos is to report and clarify that which has been written and not to denigrate, or belittle, what has become current knowledge over the past centuries. The reality of the immense size of the structure of the Cakravāla has been compared with those of the Earth's. There are minor controversies between King Ruang's *Three Worlds*, the *Abhidharmakośabhāṣyam of Vasubandhu*, and works by Kloetzli and Sadakata on the sizes and configurations of items in the cosmos. Even though some of these are minor they will be addressed if appropriate.

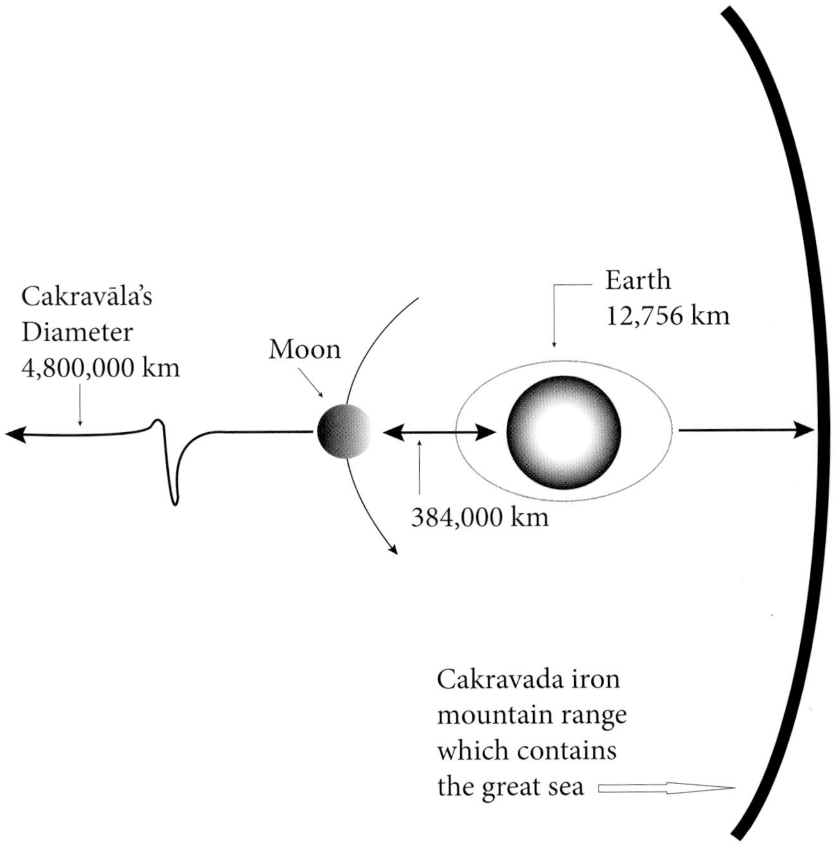

Figure 9. The diameters of the Cakravāla and the Earth[76]

The Cakravāla

The Cakravāla is a linear structure which encompasses a range of properties from hells below to heavens above. The Cakravāla <u>is a flat world.</u> At the center of the *Cakravāla* is the great soaring Mount Meru surrounded by mountains, oceans, and continents. The placement of the elements of the world is extremely orderly nothing is left to chance. Above the soaring central mountain are the multi-layered heavens. Below the mountains, continents an ocean are the segmented hells. The size of the Cakravāla, as described in the texts, is immense. Figure 9 compares the diameters of the Cakravāla and our Earth's

The diameter of the Cakravāla is 1,200,000 yojana (4,800,000 kilometers / 3,000,000 miles). The diameter of the Cakravāla is around 375 times larger than that the Earth's. To put this in perspective, Figure 9 shows the diameter of the Cakravāla compared to our world.

XIII Henry Alabaster, Esq. was interpreter for Her Majesty's consulate general in Siam.

The structure of the Cakravāla, as illustrated in Figures 10 and 11, shows a planet resting on a great circle of wind supported in turn by a mass of water, which supports a column of golden earth. This golden earth supports the mountains, the great ocean and the four continents. The great circular ocean, which exceeds any on our earth by thousands of kilometers, is contained by the Cakravada,[77] an iron mountain range encircling the planet. At the dead center of the planet is found the great axial mountain, Meru, around which all physical features are located. The continents of this cosmic planet, not unlike our earthly island continent of Australia, are at the four cardinal points. Around Mount Meru are seven gold Sattaparibhanda mountains shown in square configuration which contain the Sidantara Sea. From the outermost range of the mountains which, encircle Mount Meru to the walls of the iron mountain is a distance of 320,000 yojana[78] (1,280,000 kilometers / 800,000miles).

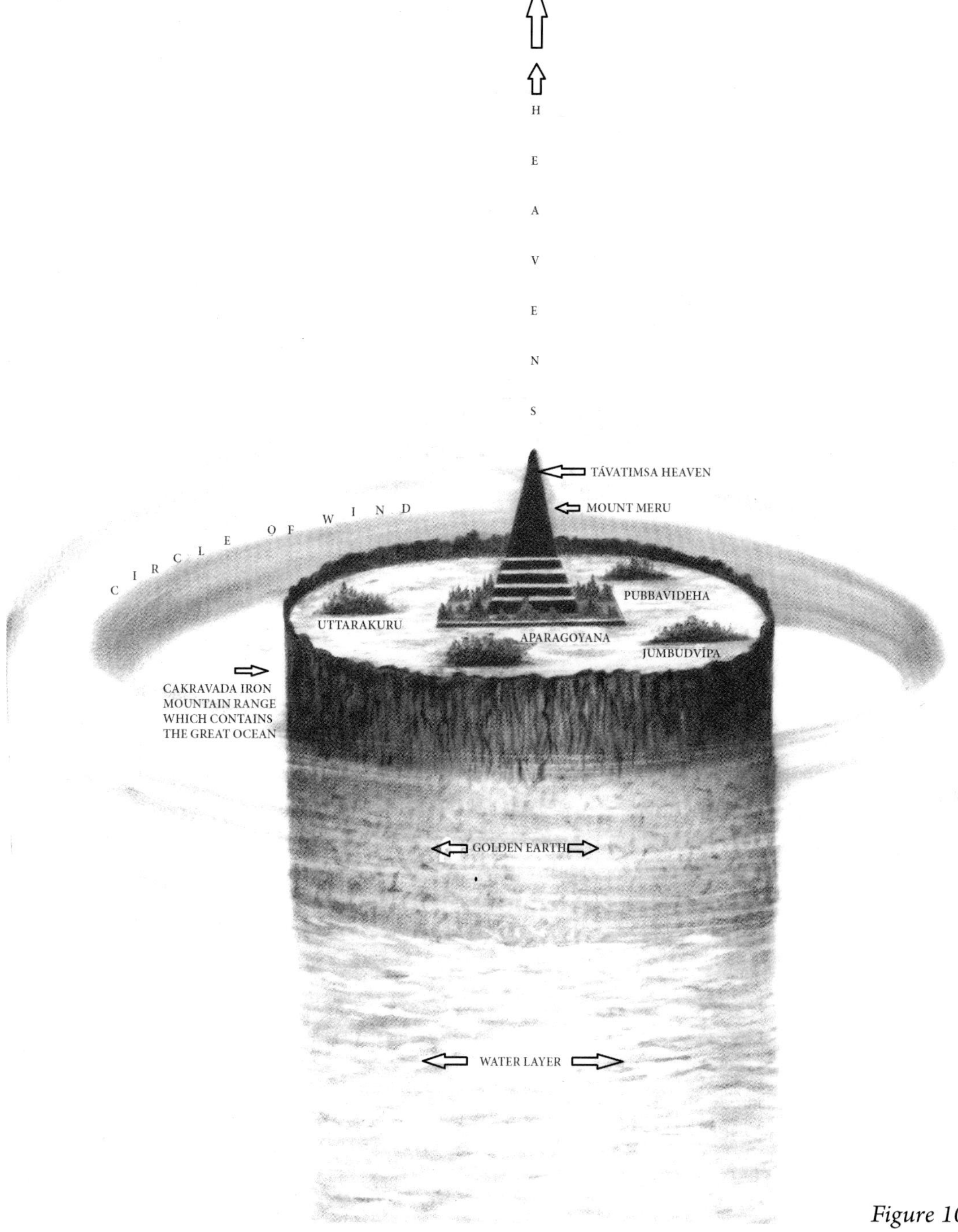

Figure 10. A Sketch of the Cakravāla

80,000 YOJANA
320,000 KILOMETERS
200,000 MILES

80,000 YOJANA
320,000 KILOMETERS
200,000 MILES

OUR EARTH TO SCALE

320,000 YOJANA
1,280,000 KILOMETERS
800,000 MILES

8 MILLION YOJANA

16 MILLION YOJANA

Figure 11. A Section through the Cakravāla

Figure 11 shows the overwhelming dimensions of the Cakravāla when compared to that of our earth.

It is thought that vast numbers of identically structured systems, Cakravālas, exist throughout the universe,[79] each resting on great circles of wind. There may be an infinite number of Mount Meru systems ranging in scale from macro to microcosmic.

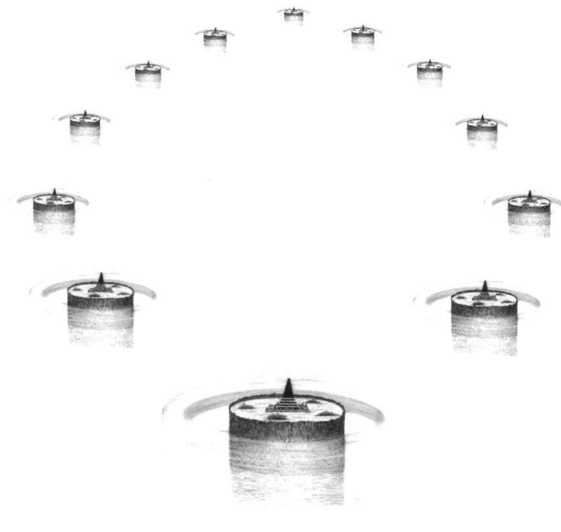

Figure 12. An interpretation of the myriads of Cakravālas *in the cosmic universe*

Plan Views

Two plan views are presented. The first, Figure 13, is a semi-scale view which shows the relative sizes and location of features. Figure 14 was extracted from the manuscript which shows the manuscript artist's impressions of the Cakravāla. Given the huge dimensions of the Cakravāla, the size of the continents shown in this figure is greatly exaggerated; the artist's main interest was to teach, not to reflect factual accuracy.

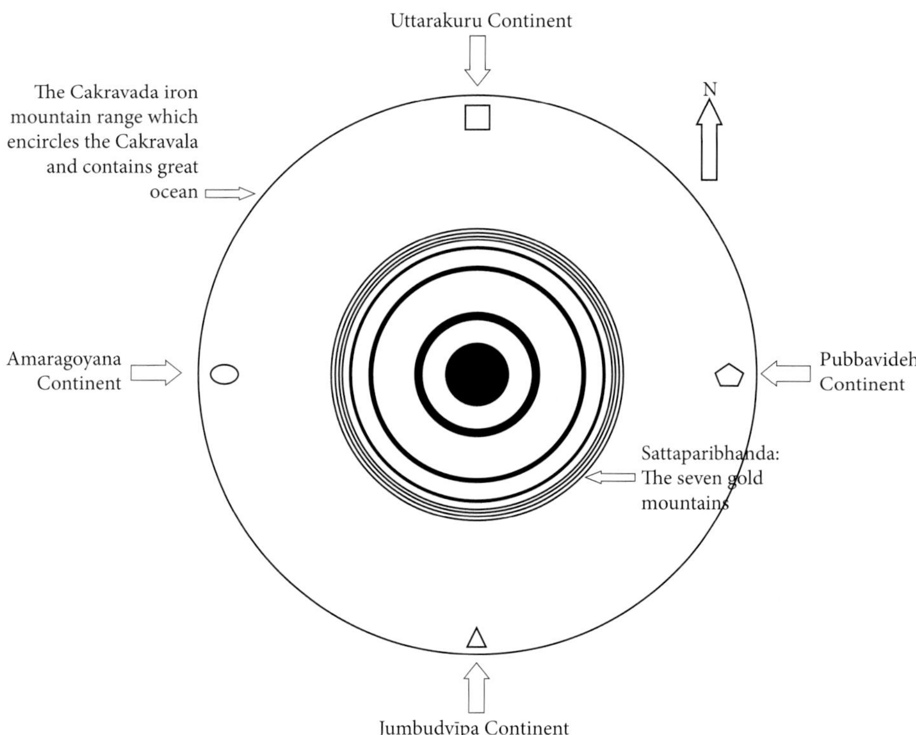

Figure 13. Plan of the Cakravāla[80]

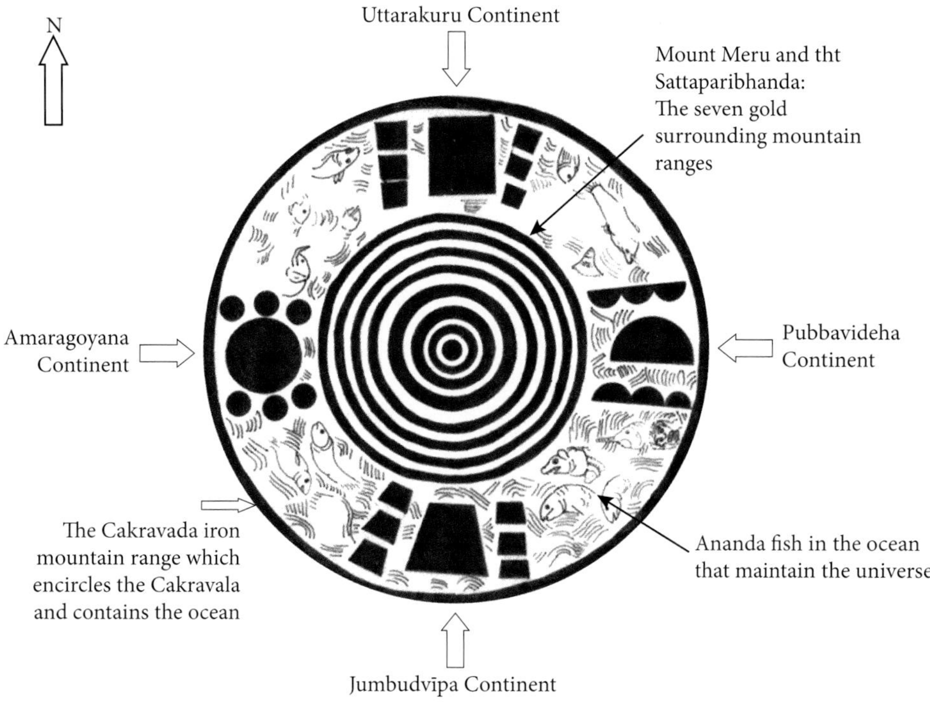

Figure 14. Manuscript Plan of the Cakravāla

Mount Meru

Mount Meru's mythical roots are from ancient India, perhaps concieved by the thought that there is a world axis and that it is Meru. According to both Hindu, Buddist and Jain beliefs Meru is located at the center of the universe and is the home of gods. This thought is not based on geographical or historical data[81] but from an ancient idea which permeated the thoughts of embronic civilizations and was put forth when writing cause ideas to coalese and consolidated. The symbolism of Mount Meru exists with the Hindus, Buddhists and Jains among others. It is expressed not only in thought but also physically in wood and stone. Meru's were built as funeral pyres for kings some of which reached well over 100 meters high.[82] In Cambodia, Angkor Wat is thought to be a replica of the cosmic universe built in stone.[83] Mount Meru may be found in Thailand at Wat Arun,[XIV] the Temple of the Dawn.

Angkor Wat

Wat Arun

Mount Meru is the great vertical cosmic *axis mundi* of Buddhist cosmology. It is the great pillar of both the structural and, spiritual worlds. Mount Meru is the focus and center of a very orderly world. Seated sedately dead center in the middle of the Cakravāla, it is surrounded by the annular ranges of seven golden mountains then, a great ocean in which serenely rest the four continents and their islands; all of which is contained by a great iron wall, the Cakravada Mountain Range. This range delineates the limits of the world and contains the waters of the great ocean. The size and configuration of Mount Meru varies slightly in size and detail depending on source. The oldest, dated to the first centuries of the Common Era, comes from the *Abhidharmakośabhāṣyam of Vasubandhu.* It describes a slightly smaller and different Mount Meru than that described in the Reynolds and Reynolds translation of King Ruang's Trai Phum.[84] These differences are shown overleaf, and described in the two figures.

The *Abhidharmakośabhāṣyam* has Mount Meru soaring from its base 80,000 yojanas deep in the ocean, to 80,000 yojanas (320,000 kilometers) into the sky. Atop

XIV A recent major study of the interface of Wat Arun and Mount Meru maybe be found in Chatri Prakitnonthakan's. *The Philosophical Construct of Wat Arun.*

the summit is the Tāvatimsa Heaven, the home of the 33 devas, is captained by the deity Sakka. The summit is square and the bounds of Tāvatimsa Heaven are 20,000 yojanas[XV] on each side.[85]

The mountain is layered with four terraces, each equally distant from the other. The first terrace is for the Yakshas; the second for the wearers of the crown; the third is for the 'always intoxicated'; and the fourth is for the four kings of the cardinal quarters with their attendants.[86] The terraces also have other appellations. The first one can be called 'Pitcher in Hand'; the second the 'Bearer of Garlands'; the third, 'Always Ecstatic'.[87]

The Trai Phum has a larger mountain springing from a depth of 84,000 yojana in the Sidantara Sea to a height of 84,000 yojana above the sea. It is 84,000 yojana thick.[88]

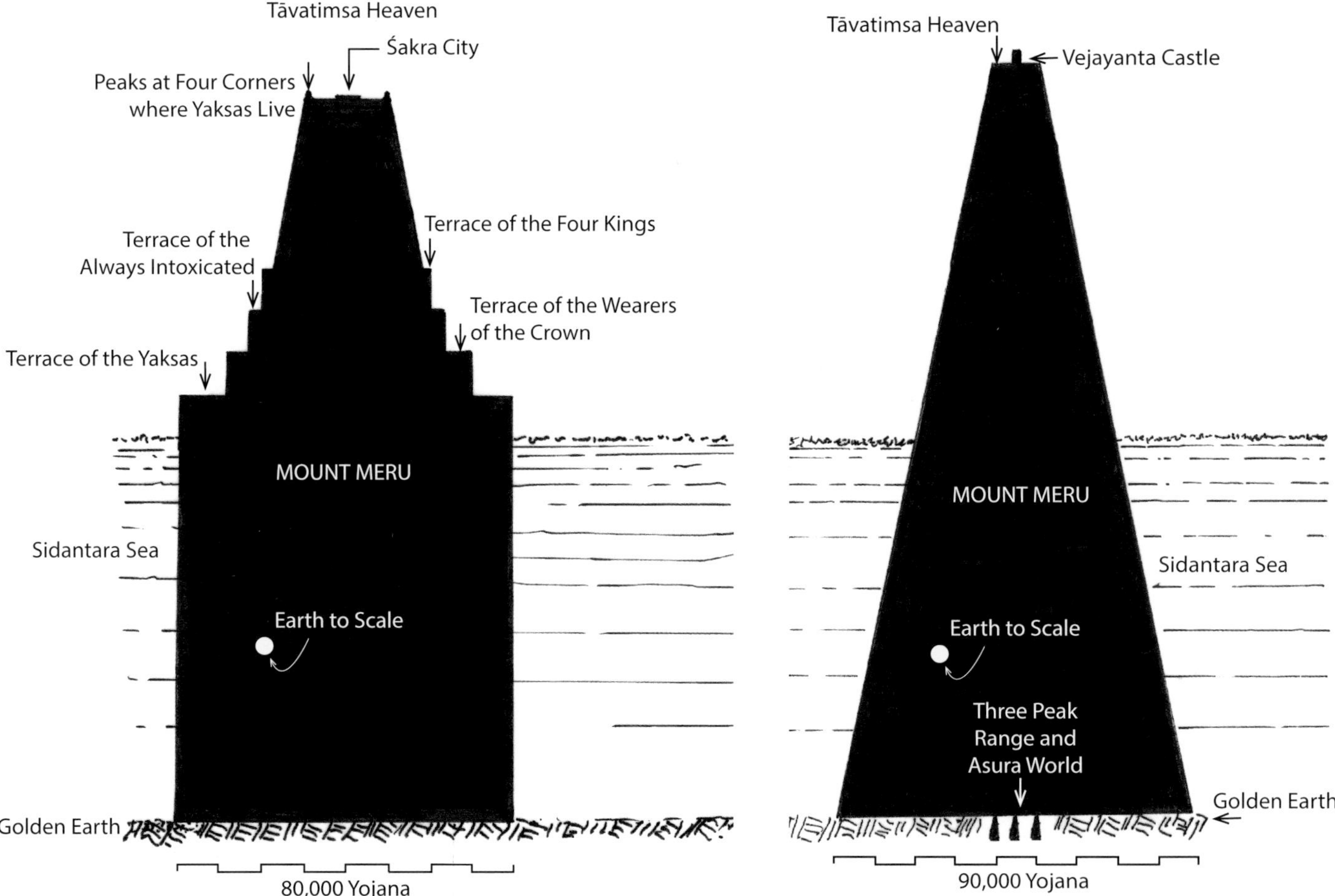

Figure 15. Configuration of Mount Meru according to the Abhidharmakośabhāṣyam

Figure 16. Configuration of Mount Meru according to King Ruang

The mountain's four quarters are colored as follows: north, gold; south, sapphire gems; east, silver; west, crystal gems.[89] Tāvatimsa Heaven is 10,000 yojana across.[90] At the base of Meru three mountains, each 4,000 yojana high, support Meru and form

XV The size of Tāvatimsa Heaven is given by various accounts at either 10,000, 20,000 or 80,000 yojanas on a side. However, if it were 80,000 yojanas, that would be the same size as Mount Meru which would result in a very large and straight column.

a place for the Asura world. Vejayanta castle[XVI] in the center of Tāvatimsa heaven is the home of Indra. The castle is 25,600,000 wa high (3,200 yojana/12,800 km) and covered with gems[91] is located in Śakra City.[XVII]

Sattaparibhanda Golden Mountains and the Sīdantara Sea

The seven gold annular (circular) mountain ranges, encompassing Mount Meru, are known as Sattaparibhanda. The following tabulation identifies these ranges and their interdigitating seas and the decrease in the dimensions of mountain ranges and seas as they move away from the center of Mount Meru.

Mount Meru – 80,000 yojana thick
Sīdantara Sea 80,000 yojana wide and 80,000 yojana deep.

Yugandhara Mountain Range – 40,000 yojana high and thick
Sīdantara Sea 40,000 yojana wide and 40,000 yojana deep.

Īsadhara Mountain Range – 20,000 yojana high and thick
Sīdantara Sea 20,000 yojana wide and 20,000 yojana deep.

Karavika Mountain Range – 10,000 yojana high and thick
Sīdantara Sea 10,000 yojana wide and 10,000 yojana deep.

Sudassana Mountain Range – 5,000 yojana high and thick
Sīdantara Sea 5,000 yojana wide and 5,000 yojana deep.

Meemindhara Mountain Range – 2,500 yojana high and thick
Sīdantara Sea 2,500 yojana wide and 2,500 yojana deep.

Vinataka Mountain Range – 1,250 yojana high and thick
Sīdantara Sea 1,250 yojana wide and 1,250 yojana deep.

Assakaṇṇa Mountain Range – 625 yojana high and thick

Outside the Assakanna mountain range lies the great ocean and the four continents. Cakravāḍa, the perimeter iron mountain range, lies some 320,000 yojanas away.[92]

Figure 17 presents a cross section of the Sattaparibhanda Mountains and the intervening seas to explain how the transition to shallower waters is made. The artists who drew the Sattaparibhanda Mountains in the manuscript had a simple way to present it. As shown in Figure 18 the sea bottom is cantilevered off the great mass of Mount Meru. The top of Mount Meru is flat to accommodate Tāvatimsa Heaven 20,000 yojana on each side.

XVI Also called Vaijayanta, the divine palace of Indra (Lithai 1985, p. 303).
XVII Also called City of Traitrimsha (Lithai 1985, p. 303).

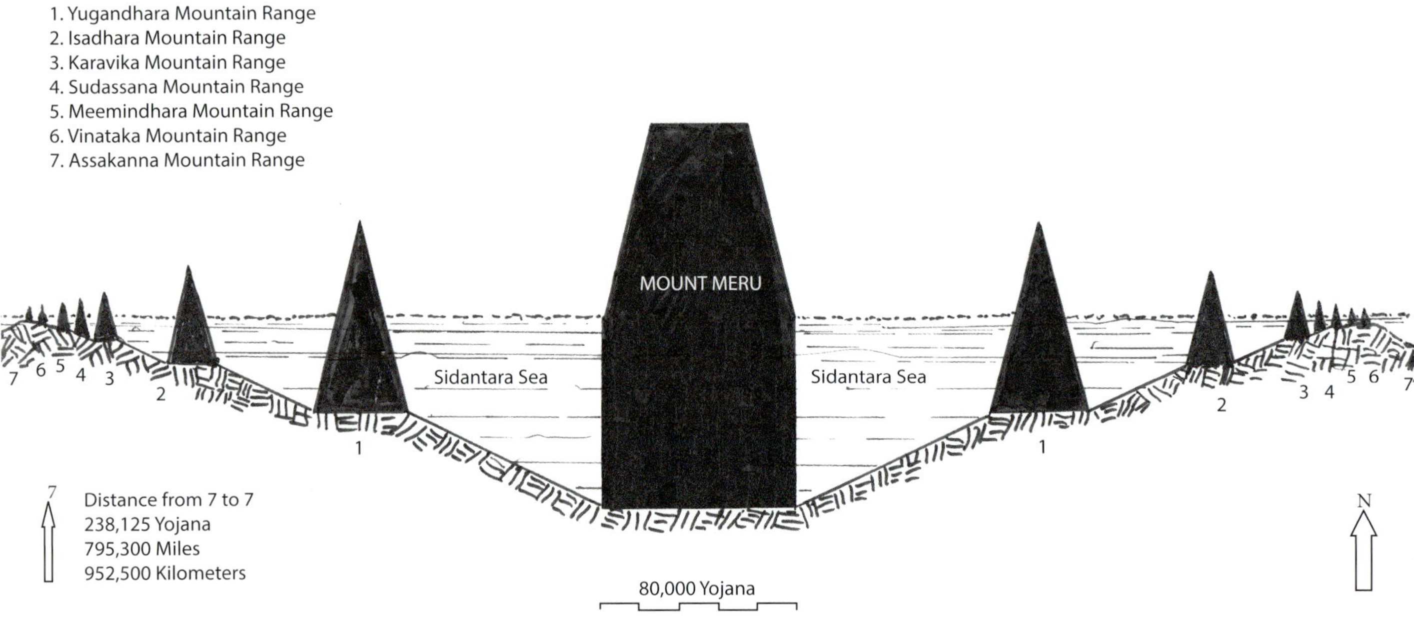

Figure 17. Cross Section of the Sattaparibhanda Mountains

Figure 18 abstracted from the manuscript shows impressive innovation in presenting the same configuration as presented in Figure 12 of the mountains receding in size and of the increasing shallow depth in the Sidantara seas. Similar illustration is to be found in most Theravāda manuscripts which deal with the cosmos.

The distances are awesome, as is everything else in the Theravāda cosmos. From the walls of Mount Meru past the surrounding mountain ranges and seas to the great ocean, is 278,125 yojana (or 1,112,500 km./695,000 miles) The configuration of the mountains and seas around Mount Meru can, as indicated, from the manuscript also be circular. King Ruang suggests a circumferential configuration.[93] Akira Sadakata in *'Buddhist Cosmology'*, makes a case for a square configuration for the mountains and seas surrounding Mount Meru.[94] Donald S. Lopez Jr. in his book *The Story of Buddhism* also suggests that Mount Meru is square in configuration.[95] which may suggest square configuration of the Sattaparibhanda Mountain ranges. Both configurations are shown in Figures 19 and 20.

Figure 18. Manuscript Section of the Sattaparibhanda Mountains

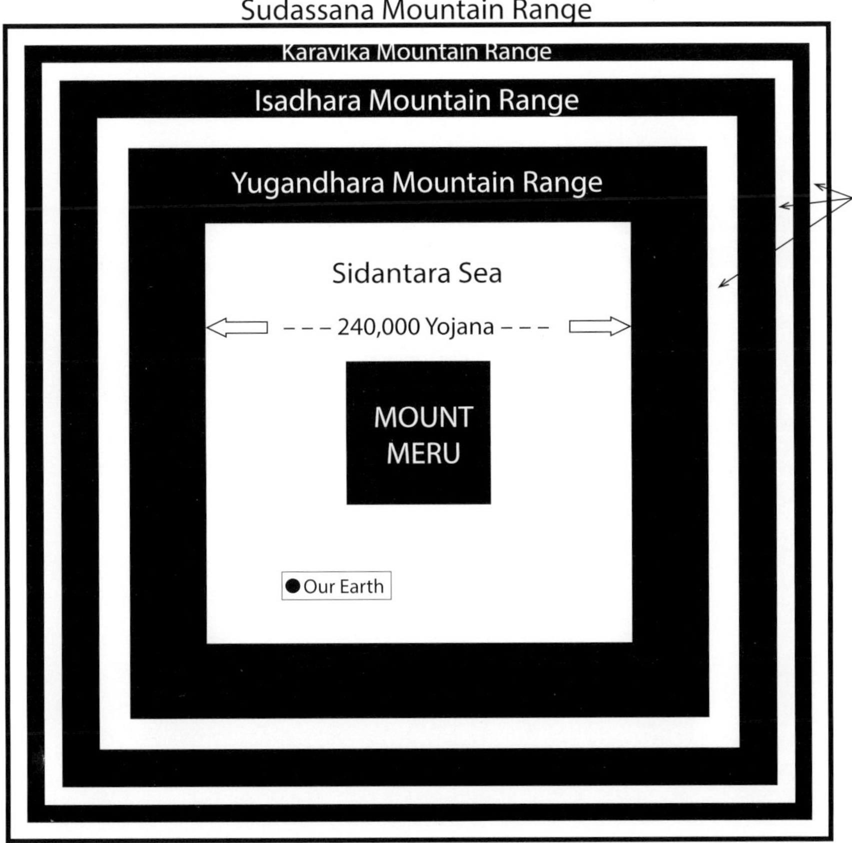

Figure 19. Square configuration of the first four Sattaparibhanda Mountain Ranges

Because of difficulties with delineation this sketch can only show the first four mountain ranges and the seas surrounding Mount Meru to scale. The box (inset) contains a black dot in a text box which approximates the earth to scale. Earth's diameter of 3,185 yojanas is approximately 1/25th of the width of Mount Meru[96]

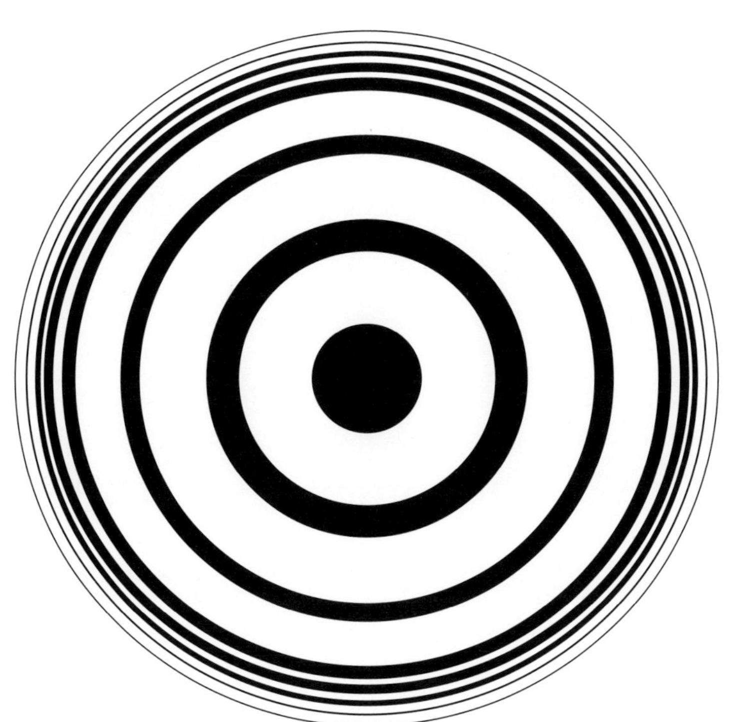

Figure 20. The Sattaparibhanda Mountain Ranges – annular configuration[97]

Table 3 summarizes the size and condition of these great insular land masses and their inhabitants.

TABLE 3. THE FOUR CONTINENTS, Their Inhabitants and Details

Name of Continent	Location with respect to Mount Meru	Shape	Size of continents in square yoja-nas/square kilometers	Comparison of cosmos continents with Australia (7,686,850 km/sq.)[98]	Inhabitants	Comments
UTTARAKURU	North	Square	4.000,000/ 64,000,000	8.3 x larger	Human lifespan is 1,000 years. People average 15 meters tall.	People are very wealthy and do not work. Food grows by itself. Cities are built in the air.
JUMBUDVĪPA (The Rose Apple Island)	South	Cart	1,625,000/ 26,000,000	3.4 x larger	Human lifespan is our normal span. Inhabitants are within the same size range as found on Earth today.	Our own familiar type of world.
PUBBAVIDEHA	East	Half Moon	2,453,125 39,250,000	5.1 x larger	Human lifespan is 250 years. People average 4 meters tall.	
APARAGOYĀNA	West	Round	4.906,250/ 78,500,000	10.2 x larger	Human lifespan is 500 years. People average 7 meters tall.	Inhabitants sleep on the ground.

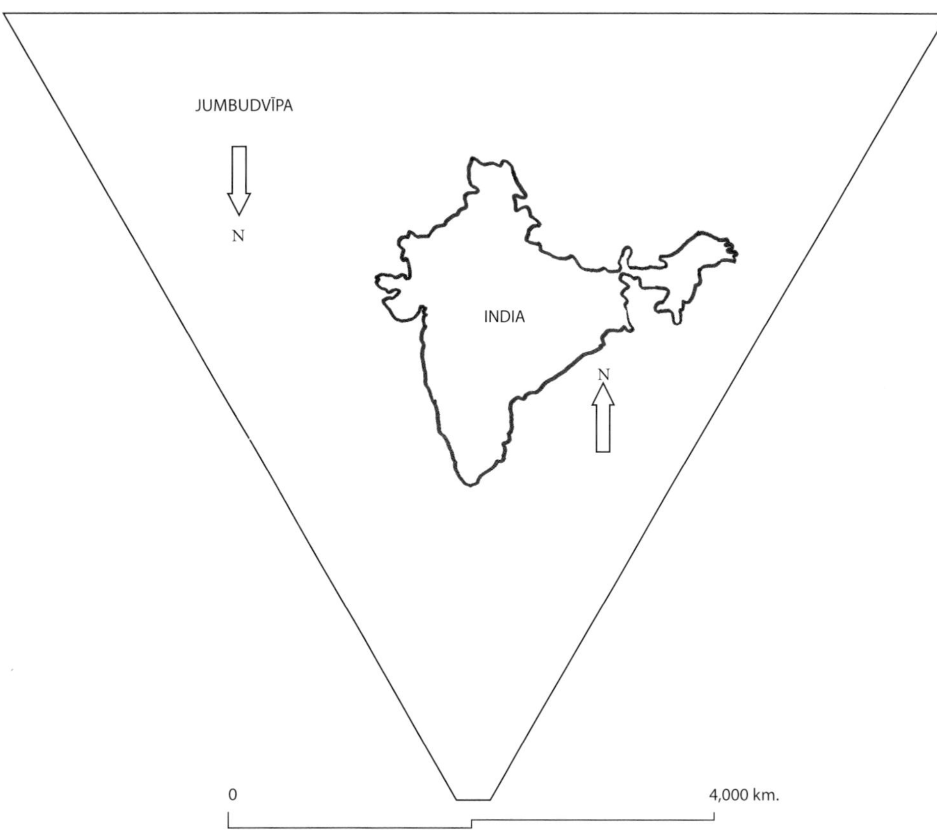

Figure 21. India and Jumbudvīpa to scale

Jumbudvīpa — *Our World*

The continent that dominates our cosmic world is Jumbudvīpa. This is the continent of our kind of humans. The continent's land mass is in the form of a slightly truncated equilateral triangle with three sides 2,000 yojanas (8,000 kilometers/5,000 miles) long with the fourth side only 3.5 yojanas (14 kilometers/8.75 miles) long. The configuration and size was derived from the *Abhidharmakośabhāṣyam of Vasubandhu*[99] It should be noted that the writings of King Lithai first suggest a round shape and an east-west size of 10,000 yojana then later suggested a triangular shape.[100] The *Abhidharmakośabhāṣyam* shape, indeed, appears to be based on the geography of the Indian sub-continent. Akira Sadakata supports this view and mentions other geological features and place names which bear resemblance to those found in India.[101]

Figure 21 shows the Indian sub-continent and Jumbudvīpa at the same scale. The orientation of both has been reversed to show configurational similarities. Jumbudvīpa is a land of great geological diversity with a *Himavanta* mountainous region of 84,000 peaks, some of which reach heights of 500 yojana (2,000 kilometers). There are great trees, elephant size birds, and great rivers that have their source in the great Anotatta Lake which flows into the sea. Surrounding the lake are mountains and caves, some of which are the homes to Pacceka-Buddhas. Jumbudvīpa is also the place where the 'maiden-fruit' tree is to be found. There are a number of immense forest areas. One of the rivers flowing from Anotatta Lake, the Ganges, undergoes a tortuous route and then splits into five rivers and five clear beautiful lakes which eventually flow to the sea. Between the rivers and lakes are the inhabitants and wondrous fruits and vegetables. Jumbudvīpa also hosts great groups of elephants led by a king elephant who flies. Jumbudvīpa has cities of happiness and joy, and ones of gold. Governance is vested in a number of city-states.[102]

Destruction and Recreation

The destruction and recreation of the Cakravāla is discussed in detail in the *Abhidharmakośabhāṣyam of Vasubandhu;* A Buddhist text written by Vasubandhu, an Indian Buddhist and philosopher circa 319–396 C.E., and also in King Lithai's 14th C.E. Traiphum. Alluded to in the introductory section of this part of the book and elsewhere, is the similarity these ancient texts have with today's scientific knowledge of similar happenings in the universe. The eventual destruction and recreation of the Cakravāla is an integral part of the Buddhist cosmos. The decay, destruction and recreation of the world systems are cyclic. Periodically, the Cakravālas in the universe are either destroyed by water, fire or wind and then after a period of great time recreated.

Summation

The foregoing, brief by design is a fusion of a number of serious sources meant to set the stage for understanding the Theravāda cosmos presented in the manuscript. It does not describe the many great details that crowd the pages of the Three *Worlds According To King Ruang* and the *Traibhumikatha: The Story Of The Three Planes Of Existence*.

PART 2

The Manuscript: An Overview

Origins

This book was initiated by the acquisition of a folding Burmese manuscript (*parabaik*) from the 19th century, which describes and illustrates the Theravāda Buddhist cosmology. The text is written in Burmese script with Pali words that are 'lifted' and distributed throughout the manuscript.[103] The material for this manuscript is native paper made of mulberry tree bark. The manuscript was created during the reign of King Tharrawddy Min, who reigned from 1837 to 1846 C.E. There is a colophon on page 2 of the manuscript which reads as follows:

'This Lawkakoonchar white parabaik was composed in the year 1204, by the good merit of Upazin Wizareintasaryi' (Translation)

Of interest is the following:

- Law-ka-koon-char is the Burmese name for cosmology manuscripts,[XVIII]
- White Parabaik refers to the format of the manuscript – a parabaik, a paper folding type manuscript coated with a white primer.

XVIII Lawka= living world and Koon-char = network and it can be roughly translated as: the network of the living world, the network of life, or the network of the cycle of life.

- Burmese year 1204 equates to 1842 C.E.[XIX]
- Upazin indicates that the donor was an ordained monk who made merit by the donation of this manuscript.

Patricia Herbert[XX] noted that Burmese cosmology manuscripts do not include a colophon with date, donor and other information,[104] which makes this manuscript rare and unique. She also noted that surviving *parabaiks* belong to the Konbaung dynasty 1752–1885 and probably from upper Burma.[105] The folded manuscript dimensions are 21 cm / 8.25" wide x 53 cm / 20.5" long x 2.5 cm / 1.0" thick, containing 46 leaves of paper, most of which contain illustrations or text on both sides. The pages are contained by two thicker, harder, stronger, covers which have illustrative material and text on their other side.

With the passage of time the manuscript has evidently been subjected to poor handling and has been damage by water obscuring the script in places and causing transference of ghost images from one leaf to another while wet. There are also a few places where corrections were made with 'white-out' type paint and written over.[XXI]

Burmese Manuscripts

Burmese manuscripts take three forms:
The first, and the most ubiquitous, are manuscripts that are carefully incised with a metal stylus onto the surface of toddy palm, or *corpha* leaves. For the most part these are concerned with religious Pali texts and are held together with strings running through holes in the outer edges. There are also well-illustrated cosmology palm-leaf manuscripts, as well as palm-leaf manuscripts that deal with more mundane subjects such as fortune telling, legal matters, literary concerns and history.

The second type of manuscript is a highly ornamental variety made of palm leaves. This includes the *kammavaca*,[XXII] with texts from the Vinaya, the rules of conduct for monks, and passages that relate to the ordination procedures for monks. This manuscript type has gilded lacquer surfaces written in heavy black script.[106]

The third manuscript type is the folding, concertina-type manuscript – the *parabaik* as presented here. *Parabaik* have the following characteristics:

- They were almost always made in the royal ateliers and, if illustrated, normally required a number of artists for a single book.[107]
- The paper was specially prepared and coated with white primer. Cosmology *parabaiks* were executed in a vertical format, as is the case here.
- As noted, Parabaiks, especially cosmological ones, were rarely signed or dated;[108]
- Burmese manuscripts normally omit the realm of Preta[109]

XIX The Burmese and Gregorian calendars differ by 638 years, therefore 1204 + 638 = 1842 C. E.

XX The former Curator of the Southeast Asia Collections of the British Library and authority on Burmese manuscripts.

XXI This type of correction may be found in Thai manuscripts. See *Trai Phum Book: Ayutthaya Manuscripts–Thonburi Manuscripts*. 3 Volumes. Amrin Printing & Publishing Company, Bangkok, 1999 (Thai language): Book published in honor of King Bhumipol's 6th Cycle, i.e. 72nd year.

XXII Pali term for passages from the Buddhist texts on the rituals and conduct of monks.

Burmese-Thai Intersects

The relationship between Thai and Burmese cosmology is intertwined, and strong. There is little question that derivative for this manuscript may have roots in the former Siamese capital of Ayutthaya, which fell to the Burmese and was sacked in April, 1767 – just 75 years before this manuscript was produced. While it is known that Thai artists were taken back to Burma with the Burmese army, and may have influenced the character and productions of manuscripts, as of today, there are no proven linkages. Patricia Herbert indicates a number of differences between Thai and Burmese cosmology manuscripts.[110] However, Derick Garnier when discussing Ayutthaya notes:

> Some things were not destroyed. In particular, the Burmese were looking for skilled artisans. Ayutthaya had been a rich and tranquil society. Craftsmanship had flourished and the arts were patronized. Burma was in comparison much less settled with neither the wealth nor time for the refinements of life.[111]

The writings of the Thai King Lithai[112] and the 'Trai Phum'[XXIII] were known to the Burmese and Cambodians.[XXIV] In this analysis of the manuscript, Thai models and sources are used to help with the explanation. The artists and scholars who painted and illustrated this manuscript followed models; perhaps some were Thai models from manuscripts that may have been seized during the sacking of Ayutthaya.

It is known that artists from Burma, Cambodia and Thailand developed similar iconographies and presentation techniques, especially for 19th century works of art.[113]

A Caveat

In the case of this particular manuscript, a number of minor conflicts exist between sources. The translations, the product of a number of hands, are one source of concern. Another is that this is a Burmese manuscript, with possibly a different interpretation of Phra Lithai's work from other books that describe the Thai Theravāda cosmos. Nomenclatures also vary slightly from source to source, especially concerning the names of the realms and levels of existence.

Iconography

A word has to be said about the imaginative presentation of the cosmos by 18th- and 19th-century Burmese (and Thai) painters. The subject the painters attempt to illustrate is complex. The scale of the Buddhist cosmos by today's standards of knowledge, provided by satellites and space exploration, is unworldly, complicated and huge. The Buddhist cosmos has been successfully approached in a number of cultures and by many artists; all have succeeded in their own methods of presentation. There also has been a number of three-dimensional presentations (including the contemporary ones to be found in this book) that may provide an added layer of understanding. In this manuscript depictions of the Buddhist cosmos are presented in two dimensional forms

---

XXIII The Tri Phum evidently has exhibited long roots in Thai Buddhism as one note the following: 'The "Traiphoon" is the standard Siamese work on Buddhist cosmogony, &c. It was compiled from presumed classical sources in A.D. 1776, by order of the Siamese King Phya Tak.' (See Alabaster 1871, p. 5.)

XXIV See the correlation established between the manuscript and the translations of the mountains surrounding Anotatta Lake in Part 4.

with a type of 'iconographic shorthand', which verges on the abstract, but is clear to the viewers. The same 'iconographic shorthand' used in this Burmese manuscript may also be found in Thai manuscripts of the same period. Some examples of this shorthand follow:

Mount Meru is shown as a very imaginative decorated column and not as described in texts as an awesome mountain of jewels. The seven surrounding mountain ranges are shown as candle-like structures and, indeed, the temple-like summits could almost be considered as flames when first viewed.

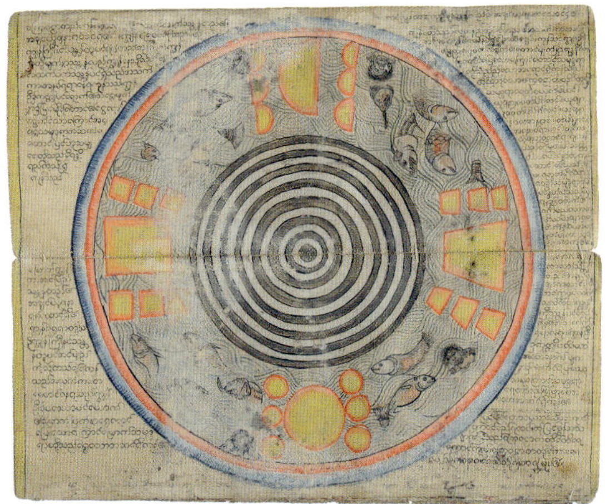

The manuscript's plan view of the cosmos is most abstract with Mount Sumeru shown as an enlarged dot and the seven surrounding mountains as circles around the dot, target-like in appearance. This approach to illustrating the world's plan view is to be found in other Thai and Burmese cosmology manuscripts.

The four continents and their satellite islands are:

Uttarakuru, north of Meru, is square;

Jumbudvīpa, south of Meru, is shaped like a cart,

Videha, east of Meru, is shaped like a half moon,

Godaniya, west of Meru, is round.

The segmented heavens are depicted as Burmese architectural style pavilions.

The segmented major hot hells are depicted as boxes full of tormented faces with flames issuing from the top of the hell.

Renderings

The paintings in this manuscript lack finesse; the robust and bold renderings of the drawings aid them to accomplish their didactic mission.[XXV] The paintings are large and take up much of the manuscript paper leaves on which they are drawn. It is assumed that a monk or a learned lay person would use the manuscript to instruct a small lay audience about the cosmos; as he talked he would unfold the manuscript to illustrate each topic.

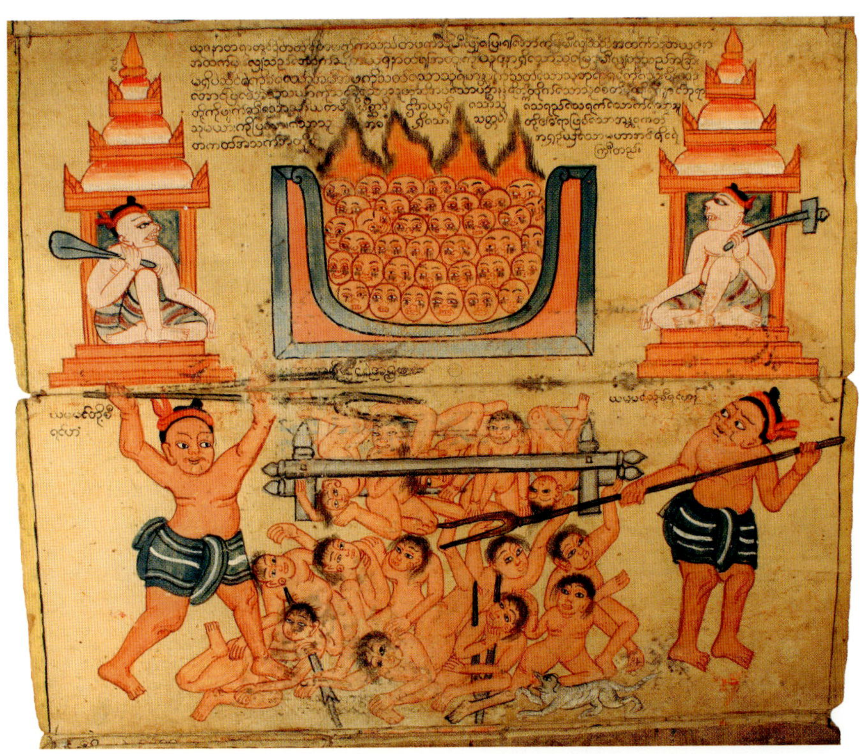

The compositional aspects of this two-leaf illustration of a hot hell illustrate the foregoing.

The upper leaf of the vignette is simplicity itself. A hot hell is flanked by two beefy, awesome, light skinned guards of the Yama king, seated in small masonry pavilions with clubs that seemingly bend under their weight. The faces of the incarcerated ones stacked on top of one another and close together, wide-eyed and running with tears, look out to the viewer, as if to warn them of the consequences of bad karma.

The lower leaf shows a number of sinners of undistinguishable gender being 'wrung-through-the-wringer' and then pierced by spears and harried by a small active dog. Large bodied, beefy, dark skinned Yama guards with human countenances, control the arrivals with double pronged spears. These guards with incipient smiles appear to enjoy their work. The message to onlookers would be quite clear; suffering in hell is not desired.

The drawings themselves are also clear. The oversized guards are shown as overwhelming when compared with the size of their charges. They are dressed with one piece *longyis*, a Burmese wrap-around lower garment, the end of which can be pulled up between the legs and fastened at the waist as a work garment. Some body definition is provided by slight changes in paint color, but otherwise the guards are two dimensional. Characteristic of Burmese and Thai iconography, the toes and fingers are of equal length.

Notes on Translations and Script

It has to be kept in mind that this manuscript is the product of a number of hands and in places it is apparent that editors and proof-readers were either lacking, or in short supply.

The translation of the manuscript proved arduous and a number of people were involved in this work (see acknowledgements). Adjustments to the translated text have been made for reasons of clarity. That being the case, in other places where there is some roughness and convolutions, but where clarity exists, the translation is presented 'as is' including questionable verb tense, has not been altered.

XXV The most notable exception to this usual characteristic of Buddhist cosmology manuscripts is the 1776 C.E. manuscript commissioned by King Taksin to ensure that the knowledge of Thai Buddhist cosmology and other religious aspects would be preserved. The drawings in this manuscript are exquisite, and micro detailed. See McGill 2005, Figures 16, 17, 102, 103, 104 and item 79 on p. 170.

The quality of the manuscript's script itself varied greatly and after translation there were places where meaning was unclear. Some of the writings in the manuscript that were translated, unfortunately, leave questions that go unanswered.

As examples:

> In Part 3, Manuscript pages 23 and 24 have interesting translated figures. Page 23 states that the lifespan in Hot Hell 5 is 8,000 years which is the equivalent of 32,000,000 human years. Page 24 states that the lifespan in Hot Hell 6 is 16,000 years which is the equivalent of 16,000,000 human years. Why the discrepancies? Perhaps these were mistakes by the scribes who should have equated 8,000 with 16,000,000 and 16,000 with 32,000,000?
>
> Manuscript pages 26 and 27 exhibit the same abnormalities as manuscript pages 23 and 24 where the human year equivalent of 'deity' years can be either 22,000 human years or 144,000 human years. There are no lifespans or human year equivalents given for Hot Hell 7 (p. 20).
>
> There is a statement in the manuscript on page 25 translated as follows: 'Thawateinthar, or 36,000,000 in the human world is equivalent to a day or night in Kalathok Hell.'

The statement floats on the page in isolation and remains unexplained.

For consistency; with the other parts of the book, the Burmese nomenclature used for the nouns have been rendered into Pali (and are shown in **bold text**). Transliterated Burmese names may be found in Appendices A and B.

When Burmese nomenclature is used, syllabication of Burmese nouns; i.e. Dok-gati- ahaik and Thu-gati-haik, favored by the translators, has been kept instead of using Dokgatiahaik and Thugatihaik.

For the most part the script has what appears to be an adjunct role to the illustrations. Upon close examination the script in the manuscript appears to be the product of several hands.

Where possible, translations have been made of the script accompanying the illustrations.

Orientation

Burmese cosmology manuscripts normally follow established precedence by being in vertical format, with the scenes running from top to bottom on both sides of the leaves, rather than opening sideways with scenes flowing from left to right or vice-versa.[114] An example of a sideways configuration as depicted in a Thai cosmology manuscript is shown in the sketch extracted and redrawn from the *Trai Phum Book: Ayutthaya Manuscripts–Thonburi Manuscripts*.[115] This manuscript's age is in excess of 200 years.[XXVI]

XXVI Burmese *parabaiks* with subjects of the Jatakas, the Ramayana, bullock-cart racing, processions of elephants, royal donations are often found in sideways presentation (see Herbert 1999, pp. 89–102).

The Manuscript's Order of Presentation

Manuscripts which deal with cosmology and the universe normally begin with the heavens and then descend level by level to the hells.[116] This manuscript follows that procedure but deviates, in that the hells are introduced in the middle. This appears to be awkward, as the 'hells' precede the illustrations of Jumbudvīpa and the Cakravāla. There is also a sizable concluding segment which deals with personages. Since the manuscript over its lifetime of 170 years has been subjected to abuse it is conceivable that it came apart and was put back together incorrectly. Questions arise for manuscript pages 19 to 22, which have the Preta realm in between pages which deal with hells. The Preta realm level 29 should proceeds the Niraya realm 31. However, the orderly presentation of the 11 two-leaf panels of the arhats and religious personages makes this seem unlikely. These reverse sides of these leaves fit into the manuscript correctly.

Figure 22 summarizes the sequential arrangement of the manuscript. Two paper leaves make one page with a central fold. In all, there are 46 leafs, constituting 45 pages on both sides with dialogue and illustrations and two covers. The *in situ* measurement performed for the length of the manuscript, and shown in Figure 22 appears to be at variance with a summation of the leafs and covers; i.e. 46 leafs x 21cm / 8.25" (9.66 meters / 31.62 feet) except for the fact that the string ties holding the leaves together are loose.

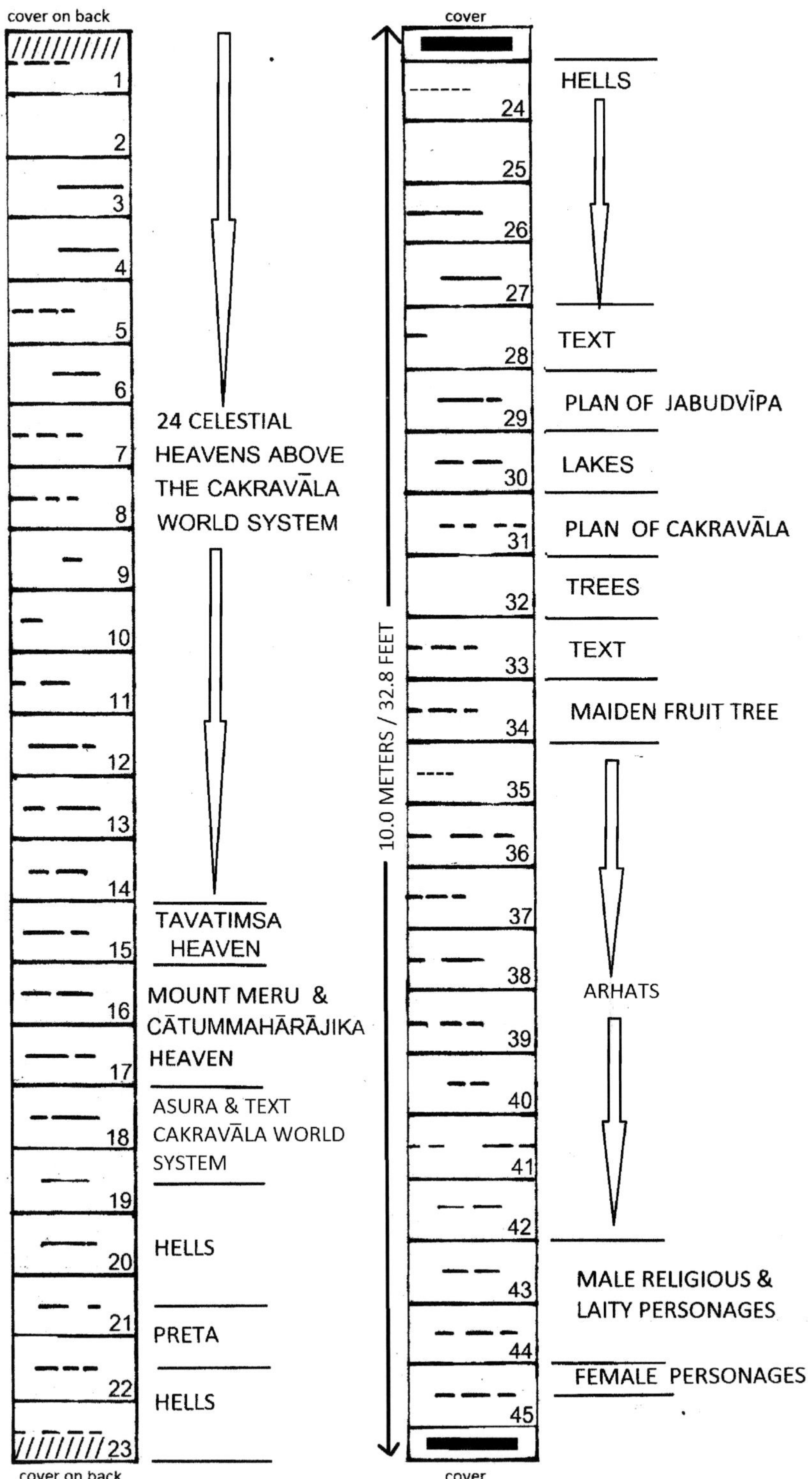

Figure 22. Sequential Arrangement of the Manuscript

Manuscript Pages in Order of Appearance

As discussed, a page consists of two leafs. There are 46 leafs with drawings and text on both sides; which constitute 45 pages and two covers. The illustrations and text of the pages can be seen interloping on each other with subsequent items intruding onto preceding pages.

1

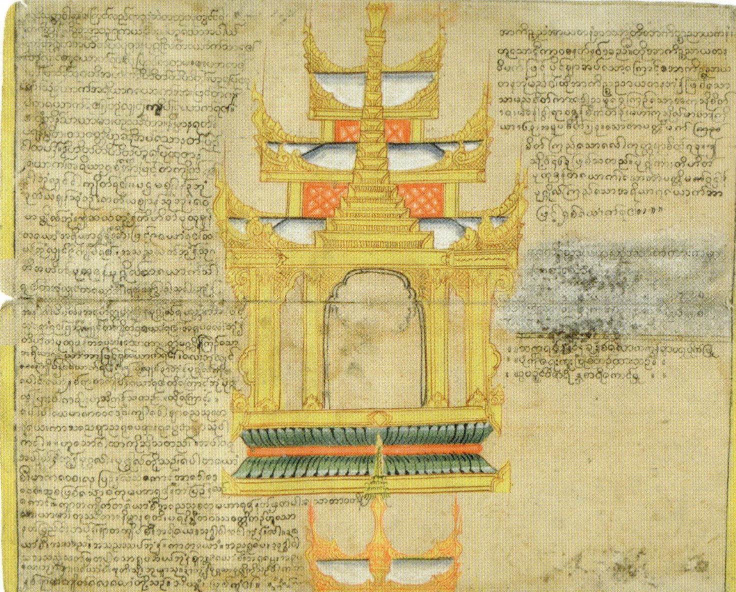

2

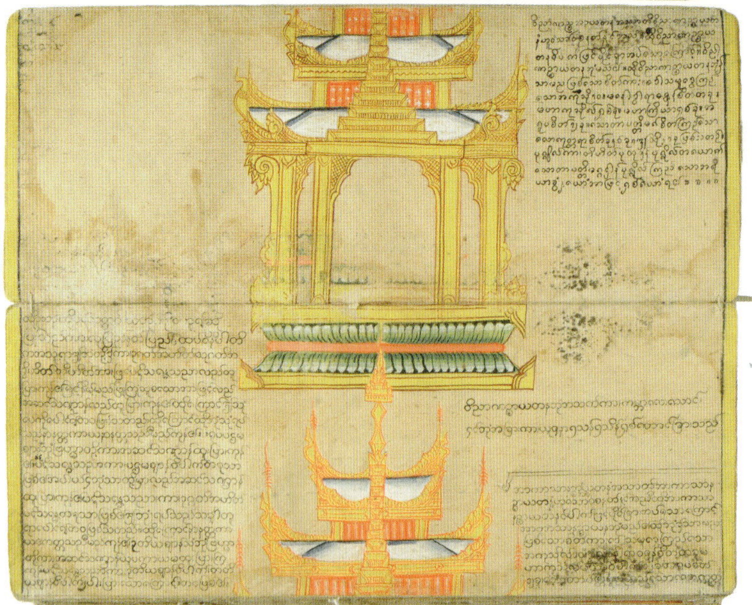

3

4

5

6

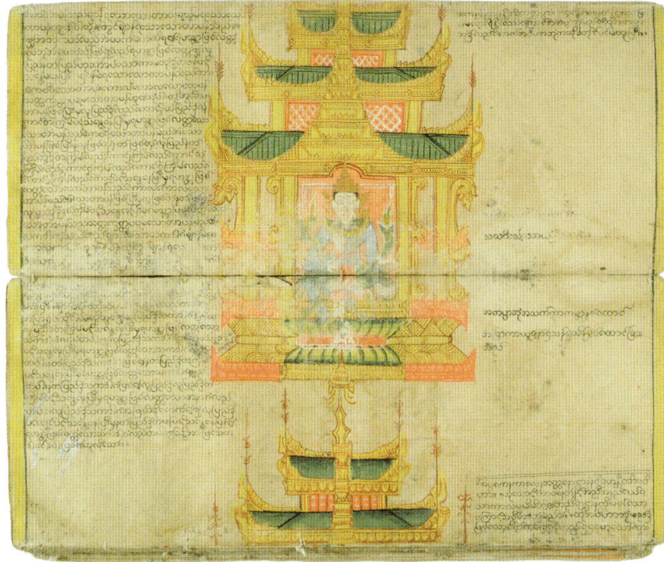

7

8

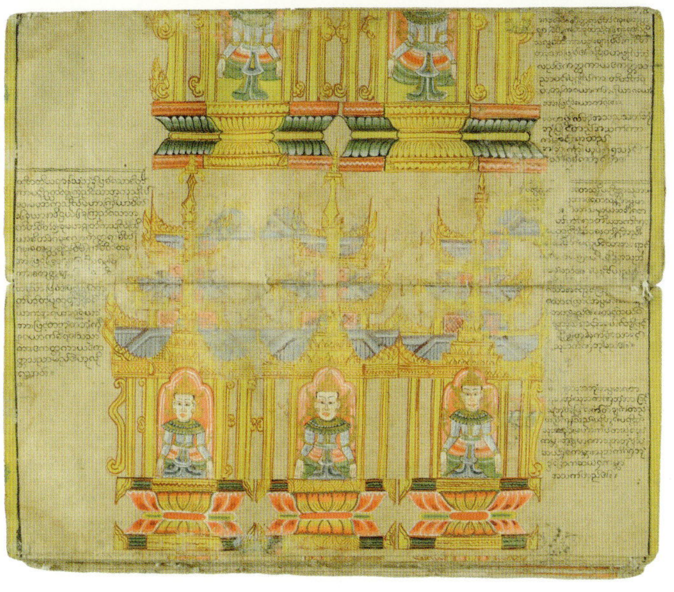

9

10

13

11

14

12

15

16

19

17

20

18

21

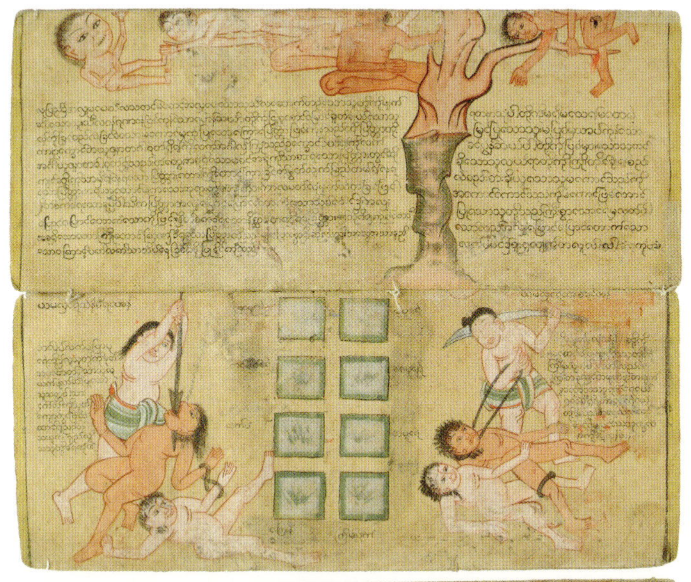

22

23

24

25

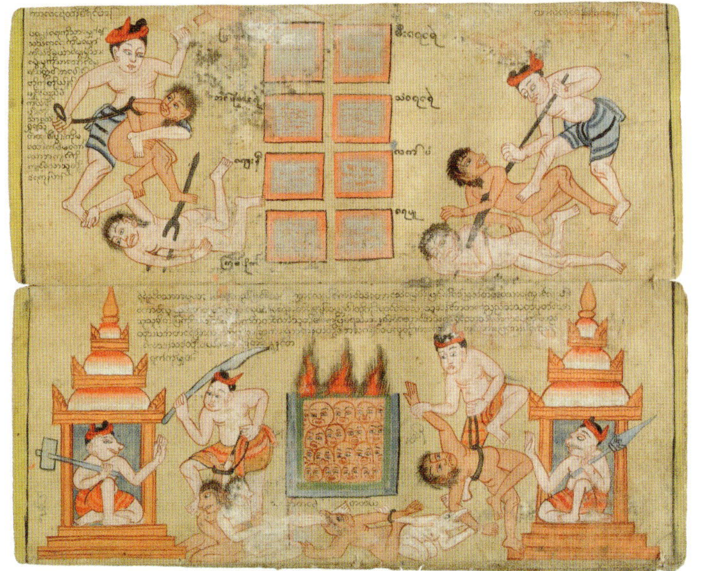

26

27

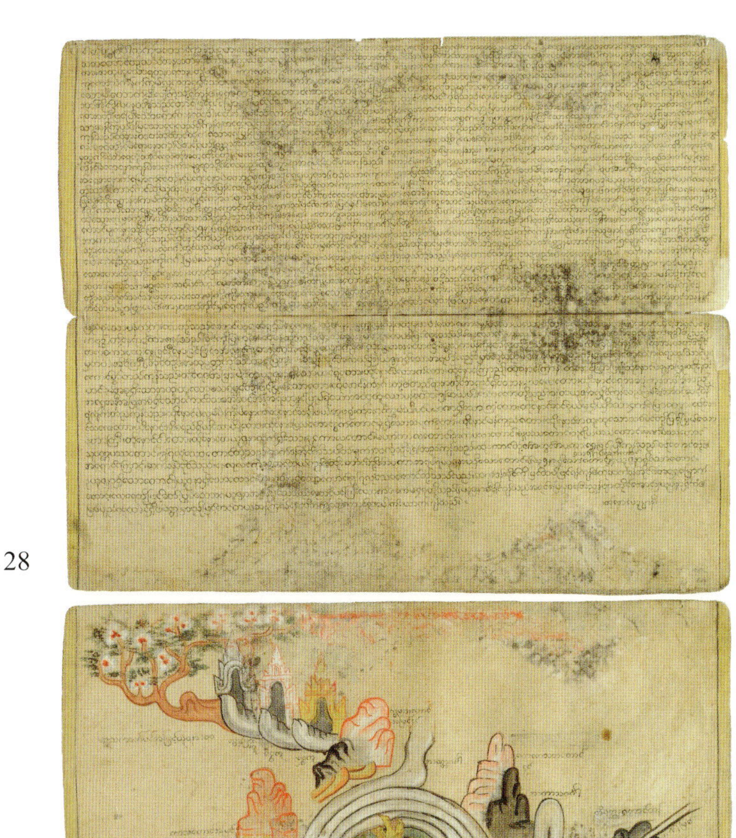

28

29

30

31

32

33

34

35

36

37

38

39

40 41 42 43 44 45

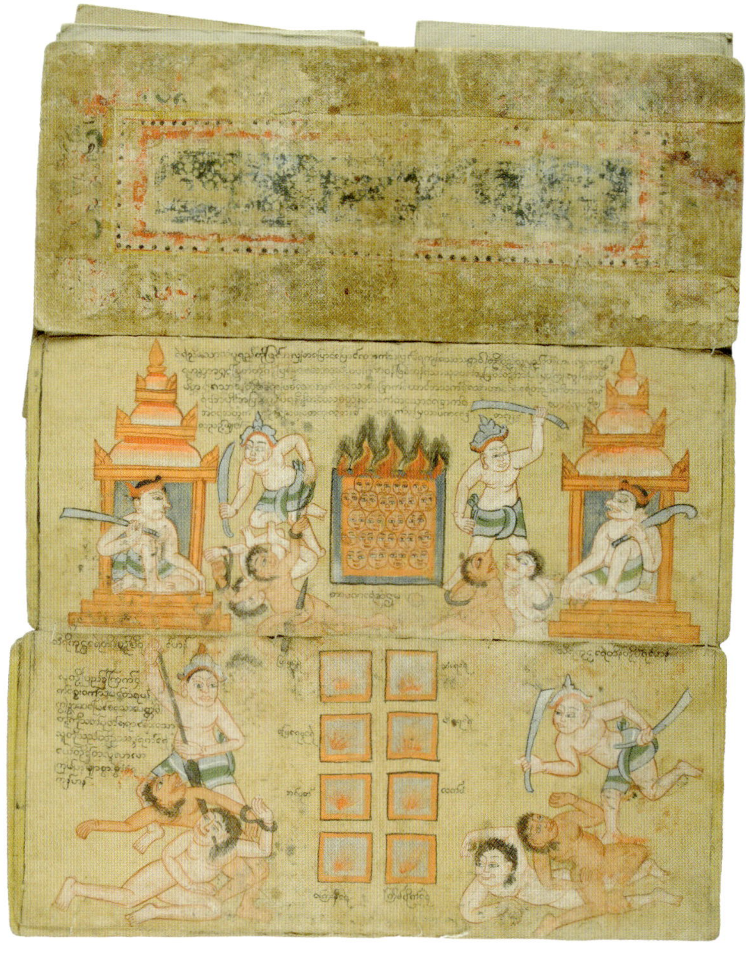

Page 1 on reverse side

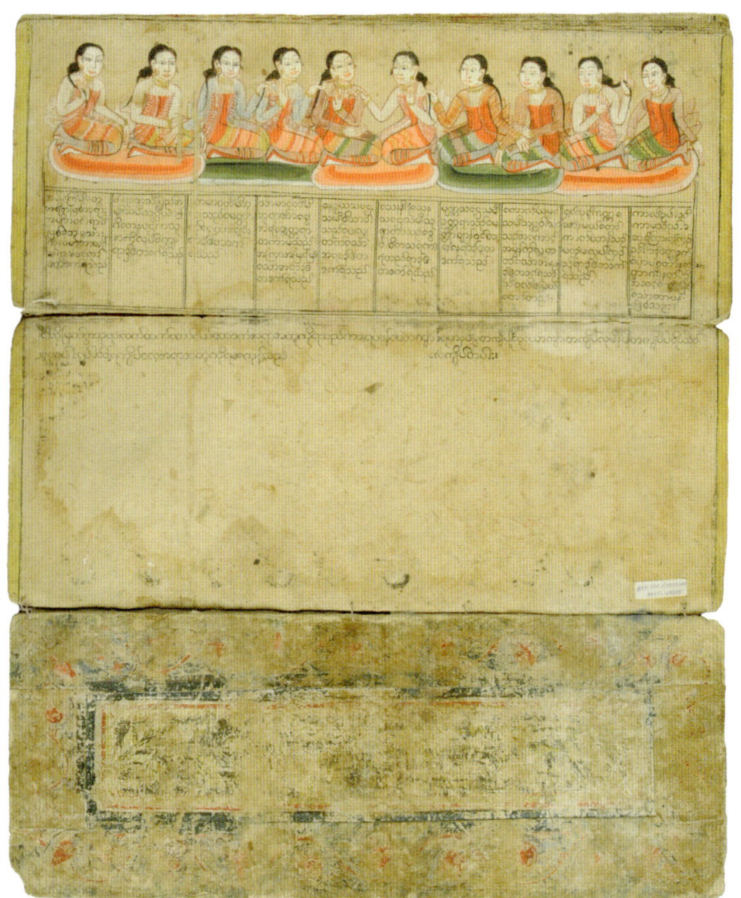

Page 23 on reverse side

PART 3

Manuscript: The Worlds of the Cosmos

The Three Worlds and 31 Levels of Existence

The Theravāda cosmos, as noted in Part One, consists of three worlds with 31 levels (or realms) into which beings can be born during their passage through Saṃsāra – the cycles of repeated births and deaths until Nibbana is obtained:

World One, Heavens without form and material factors – 4 heavenly realms

World Two, Heavens with form and material factors – 16 heavenly realms

World Three, Happiness and desire – 6 heavenly realms and 5 other realms

Table 4 provides an abbreviated listing of the heavens and other realms that were discussed in Part 1 and compares those with what is to be found in the study of the manuscript.

The presentation in this manuscript of these worlds follows that of other Thai and Burmese Theravada cosmological manuscripts[117] and begins with the many heavens and progresses though the realms of desire, failure and ends with the hells.

Partial Translation of Manuscript Page 18: The Realms[XXVII]

As noted in Part 4, for commonality with the other parts of the book, Burmese nomenclature used for the nouns have been rendered into Pali and are **shown in bold text**. The transliterated Burmese names and their Pali equivalents may be found in Appendices A and B. The presentation of the realms in the following translation follows the procedure established in King Lithai's work, in that the manuscript begins with the lower realms and proceeds to the higher ones.[XXVIII] In the manuscript's text there are four groupings of the realms, which are based on related inherent characteristics.

The outline-type format of the following translation is that of the translator, and not of the manuscript; it has been kept as it helps with clarity of understanding.

❊ ❊ ❊

The realms of all living beings are divided into four main divisions. They are: A. The four forbidden realms, B. The seven sensual realms, C. The sixteen formed realms and D. The four formless realms.

A. The four forbidden realms are (1) ***Niraya*** (the hells) (2) ***Tiracchāna*** (animals) (3) **Preta** (hungry ghosts) and (4) ***Asura***. (disposed demigods),

B. The seven sensual realms are: (1) **Manussa-** (human), (2) **Cātummahārājika**, (3) **Tāvatimsa** (4) **Yāmā-**, (5) **Tusita**, (6) **Nimmānarati-** (7) **Paranimmita-vasavattī.**

C. The sixteen formed realms are based on Jhāna levels of entry are subdivided into five stages:

   First stage Jhāna (1) **Brahmapārisajja**, (2) **Brahmaparohita**, (3) **Mahābrahmā-**

   Second stage Jhāna (4) **Parittābhā**, (5) **Appamāṇabhā**, (6) **Ābhassarā**.

   Third stage Jhāna (7) **Parittasubhā**, (8) **Appamāṇasubhā**, (9) **Subhakiṇṇā**.

   Fourth stage Jhāna (10) **Vehapphala**. (11) **Asaññasatta**.

   Fifth stage Jhāna (12) **Avihā**, (13) **Atappā**, (14) **Sudassā-**, (15) **Sudassī**, (16) **Akaniṭṭha**.

D. The four formless realms are (1) **Ākāsānaññcāyatana-**, (2) **Viññānaññcāyatana**, (3) **Ākiñcaññāyatana**, (4) **Nevasaññā-nāsaññāyatana** (T)[XXIX]

XXVII The text for the Cakravāla world system in the manuscript is to be found between the Asura, two of the hells and the Pretra realms. A question arises – why here? Why not split the text and place it where appropriate? The text translated from page 18 and which presents the realms perhaps would have been better at the beginning of the manuscript then placed between two realms, that are home to troubled inhabitants, and two hells.

XXVIII The manuscript itself begins with the heavenly realms and ends with a long inquiry into a singular realm – Niraya, the 31st and lowest level of existence, the hell world of the damned consisting of the eight great hot hells and their minor hells.

XXIX (T) = Translation.

TABLE 4. THE 33 REALMS OF THE COSMOS

Level	Pali Name[118] & Condition	Life Expectancy Mahākappa		Entry Jhāna Levels		Per Manuscript text – distance from nearest realm in yojana
	WORLD ONE WITHOUT FORM OR MATERIAL FACTORS — 4 REALMS (*Arūpaloka*)	**From Part 1 of this book**	**Per Manuscript text and illustrations**	**From Part 1 of this book**	**Per Manuscript text**[1]	
1	***Nevasaññā-nāsaññāyatana*** – Neither-perception-nor-non-perception	80,000	?	8th	none	7,508,000
2	***Ākiñcaññāyatana*** – Nothingness	60,000	?	7th	none	5,508,000
3	**Viññānaññcāyatana** – Infinite consciousness	40,000	40,000	6th	none	5,508,000
4	**Ākāsānaññcāyatana** – Infinite Space	20,000	?	5th	none	5,508,000
	WORLD TWO OF FORM AND MATERIAL FACTORS—16 REALMS (*Rūpalaka*)					
	Five Realms of Pure Abodes *(Suddhāvāsa)*					
5	**Akaniṭṭha** – The Highest of Devas	16,000	16,000	5th	5th	5,508,000
6	***Sudassī*** – Clear sighted Devas	8,000	8,000	5th	5th	5,508,000
7	***Sudassā*** – Beautiful Devas	4,000	4,000	4th	5th	
8	**Atappā** – Serene Devas without troubles	2,000	?	4th	5th	
9	***Avihā*** – Devas not falling from prosperity	1,000	1,000	4th	5th	5,508,000
	Eleven lower realms					
10	***Asaññasatta*** – Devas without perception	500	500	4th	4th	
11	***Vehapphala*** – Devas who are fruitful	500	500	4th	4th	
12	**Subhakiṇṇā** – Devas of unlimited splendor	64	64	3rd	3rd	
13	***Appamāṇasubhā*** – Devas of unlimited beauty	32	8	3rd	3rd	
14	***Parittasubhā*** – Devas of limited beauty	16	8	3rd	3rd	
15	**Ābhassarā** – Radiant Devas	8	8	2nd	2nd	
16	**Appamāṇabhā** – Devas of unlimited splendor	4	8	2nd	2nd	
17	**Parittābhā** – Devas of limited splendor	2	8	2nd	2nd	
18	***Mahābrahmā*** – The Great Brahmā	1	llegible	1st	1st	
19	***Brahmaparohita*** – Deva priests & ministers of Mahabrahmā	0.5	Illegible	1st	1st	
20	**Brahmapārisajja** – Devas in the retinue of the Mahabrahmā	0.3	Illegible	1st	1st	

	WORLD THREE OF HAPPINESS AND DESIRE, THE REMAINING SIX HEAVENLY REALMS AND FIVE OTHER REALMS (*Kāmaloka*)	Life Expectancy Celestial Years	Life Expectancy Celestial Years[2]	Life Expectancy in Human Years
	Heavenly Realms			
21	***Paranimmita-vasavattī*** – Devas who find pleasure made by others	16,000	?	
22	***Nimmānarati*** – Devas who enjoy their own works	8,000	16,000	9,216,000,000
23	***Tusita*** – The birthplace of *Bodhisattvas*	4,000	4,000	576,000,000
24	***Yāmā*** – The level of Yāmā Devas	2,000	2,000	44,000,000
25	***Tāvatimsa*** – The home of Indra (Sakka) and 32 Devas	1,000	1,000	36,000,000
26	***Cātummahārājika*** – The four kings of the cardinal quarters	500	1,000	36,000,000
	Other Realms			
27	***Manussa*** – The level of humans	Indefinite		
28	***Asura*** – The level of fallen devas and demons	Indefinite		
29	***Preta*** – The level of the hungry ghosts	Indefinite		
30	***Tiracchāna*** – The level for all animals	Indefinite		
31	***Niraya*** – The hell world of the damned	Indefinite	500	

The Heavens of the Cosmos

The heavens constitute 26 of the 31 realms and follow the order of presentation in the manuscript. The heavens are reserved for those who have done good deeds during a lifetime through the acquisition of considerable merit. Those who access the heavens after death will experience pleasure, including sensual pleasure, for extremely long periods of time. The heavens are not eternal but transitory. After a period of time, which can be very great in terms of human years, the residents in the heavens die and are re-born (see Part 6). The heavens, as with the hells and other parts of the cosmos, are segmented.

The multitude of heavens are overlaid by the associated Jhānas, which are absorbed states of mind brought about by specific states of meditative concentration and are associated with types of heaven. The manuscript has illustrations of all the heavens.

World One – 4 Heavenly Realms without Material Factors (Arūpaloka)

The inhabitants of these realms are possessed entirely of minds. They have no physical components, and are thus unable to hear Dhamma teachings.[119] Inhabitants of these four heavens reach rebirth after extremely long periods of time. These four heavens are realized after attainment in the four formless levels of meditation; the highest level of meditation on this side of Nibbana.[120] Each immaterial attainment leads to rebirth into the corresponding realm.[121]

The illustrations (manuscript pages 1, 2, 3 & 4) of these four heavens show similar ornate Burmese style princely pavilions without inhabitants, reflecting the presence of the mind, not the body. As the manuscript unfolds the pavilions are shown stacked one on another with the highest spire of the lower ones penetrating slightly into the heavenly level above.[XXX] The buildings are roofed by an ornate *phyathat;* a structured tiered roof

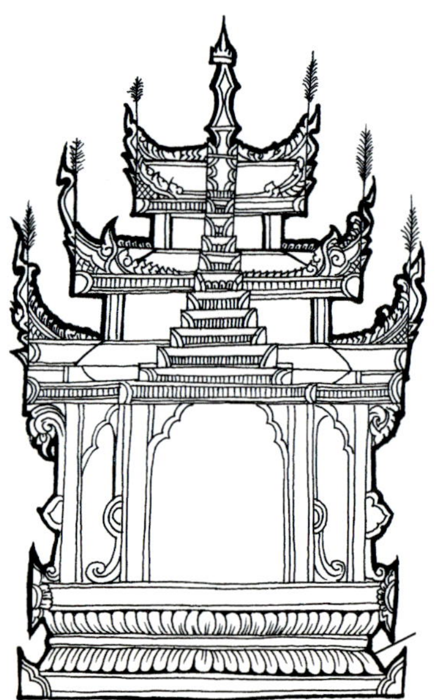

XXX The linkage, the penetration of the spire slightly into the heaven above may symbolize the desire

placed on religious, or royal buildings (see sketch below). The finials on the ends of the roofs are called *chofas,* stylized bird heads, which emerge from elaborate carved supports. Under differing terms for each culture, these architectural features are common to Theravāda religious buildings in Southeast Asia. The roof structure is supported by sturdy columns which rest on a plinth of colored lotus leaves; perhaps, in the case of the heavens these are all products of meditative processes.

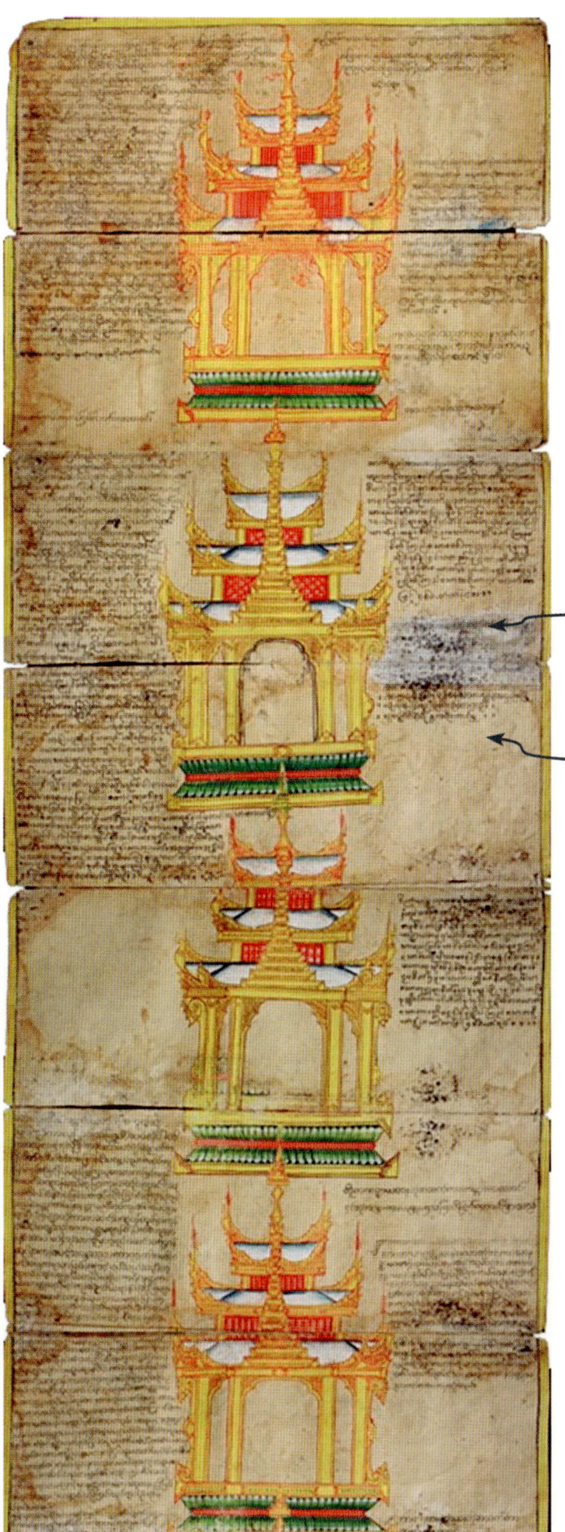

Level 1 *Nevasaññā-nāsaññāyatana* – Heaven of neither-perception-nor-non-perception

Distance from nearest realm 7,508,000 yojana and 71,856,000 from human world. (T)[XXXI]

It was named *Nevasaññā-nāsaññāyatana* because neither has, nor lacks? (T)

Jhāna entry level and lifespan: none were given per manuscript

Level 2 *Ākiñcaññāyatana* – Heaven of nothingness

Jhāna entry level and lifespan: none were given per manuscript

Note the 'white-out' area where corrections have been made and written over. This kind of correction occurs in a number of places throughout the manuscript.

Colophon: This Lawkakoonchar white parabaik was composed in the year 1204, by the good merit of Upazin Wizareintasaryi. (T)

Level 3 *Viññānaññcāyatana* – Heaven of infinite consciousness

Lifespan: Mahākappa 40,000.

Distance from nearest realm 5,508,000 yojana. (T)

Jhāna entry level: none given per manuscript.

Level 4 *Ākāsānaññcāyatana* – Heaven of infinite space

Distance from nearest realm 5,508,000 yojana. (T)

Jhāna entry level and lifespan; none given per manuscript.

of having re-birth into a higher heaven. Examination of the drawings show it to be applicable only to the first 23 levels; with Yāmā, Tāvatimsa and Cātummahārājika heavens excluded. The penetration of the spire is also shown in Level 28, the Asura realm.

XXXI (T) = translation from manuscript text.

World Two – 16 Heavenly Realms with Material Factors (Rūpaloka)

The Five Pure Abodes (*Suddhāvāsa*)

The first of the 16 realms with material factors are the Five Pure Abodes, accessible only to non-returners (*anagami*) and arhats. Beings that become non-returners in other levels, or realms, are reborn here, where they attain arhatship. Among inhabitants is the deity Sahampati who appears several times in the Pali Canon and is famous for interceding with the Buddha to overcome his hesitation on whether or not to preach the Dhamma to the world. Sahampati pleaded with him to do so, so that those 'with a little dust before their eyes' might achieve salvation.[122] In the manuscript (Manuscript Pages 4, 5, 6, 7 & 8) each of the pure abode heavens houses a Brahmā figure, wearing colorful long-sleeve attire, seated in an ornate pavilion.

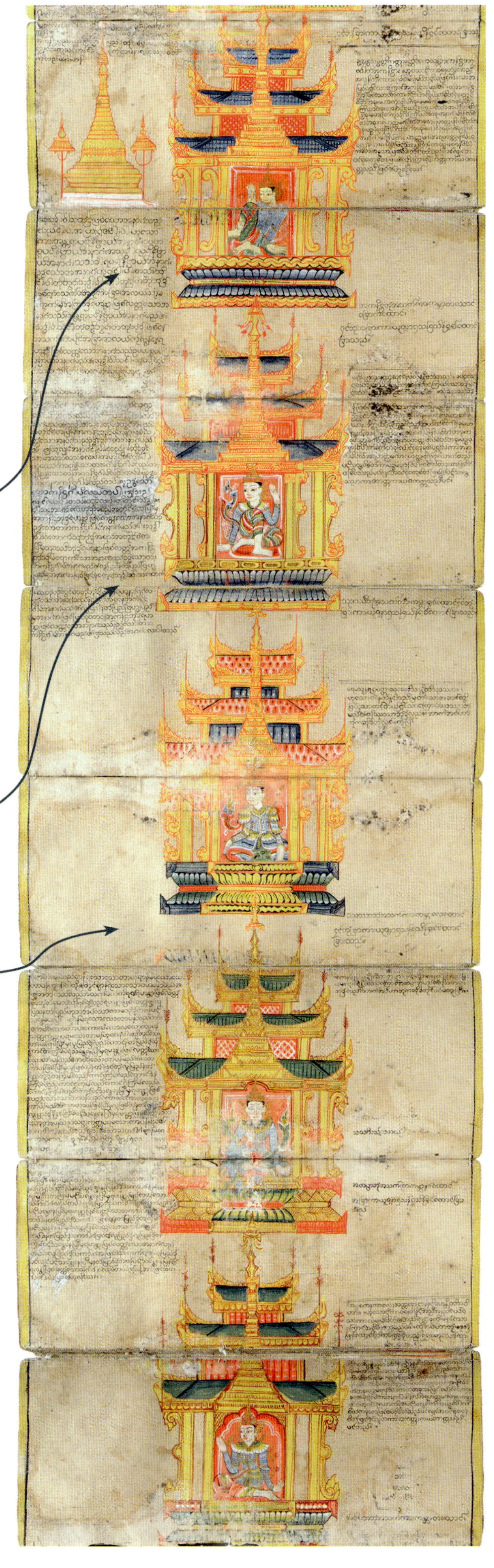

Level 5 *Akaniṭṭha* – Heaven of the Highest Devas
Lifespan: Mahākappa 16,000
Jhāna entry level: 5th
Distance from nearest realm is 5,508,000 yojana. (T)

The chedi shown in Level 5 is the Dussacetiya,[123] or Garment Chedi, which enshrined the garments Prince Siddhartha discarded when he renounced the life he had been leading and embarked on the road towards enlightenment. The Chedi is characterized by bulbous rings of the spire, a prominent architectural feature of the Ayutthaya period.

Level 6 *Sudassī* – Heaven of clear sighted Devas
Lifespan: Mahākappa 8,000
Jhāna entry level: 5th
Distance from nearest realm is 5,508,000 yojana. (T)

Level 7 *Sudassā* – Heaven of beautiful Devas
Lifespan; Mahākappa 4,000
Jhāna entry level: 5th
Distance from nearest realm is 5,508,000 yojana. (T)

Level 8 *Atappā* – Heaven of serene Devas without troubles
No lifespan figure given.
Jhāna entry level: 5th (T)
No distance from nearest realm given.

Level 9 *Avihā* – Heaven of Devas not falling from prosperity
Lifespan: Mahākappa 1,000
Jhāna entry level: 5th
Distance from nearest realm is 5,508,000 yojana. (T)

Levels 10 to 20 – Heavens with Material Factors – 11 Realms

The *Asaññasatta and Vehapphala* heavens, Levels 10 and 11, are two of the higher heavens of Brahmā where gods and men who have attained fourth Jhāna are reborn.[124] These heavens are shown beside each other as linked pavilions, each of which houses a seated Brahmā personage (manuscript pages 8 & 9) Levels 12 through 20, the nine upper domains, are shown as three linked pavilions each of which houses a Brahmā (manuscript pages 9, 10 & 11).

Level 10 ***Asaññasatta*** – Heaven of Devas without perception
Lifespan: Mahākappa 500/Jhāna entry level: 4th. (T)

Level 11 ***Vehapphala*** – Heaven of Devas who are fruitful
Lifespan: Mahākappa 500/ Jhāna entry level: 4th. (T)

Level 12 ***Subhakiṇṇā*** – Heaven of Devas of unlimited splendor
Lifespan: Mahākappa 64/ Jhāna entry level: 3rd. (T)

Level 13 ***Appamāṇasubhā*** – Heaven of Devas of unlimited beauty
Lifespan: Mahākappa 8 / Jhāna entry level: 3rd. (T)

Level 14 ***Parittasubhā*** – Heaven of Devas of limited beauty
Lifespan: Mahākappa 8/ Jhāna entry level: 3rd. (T)

Level 15 ***Ābhassarā*** – The heaven of radiant Devas
Lifespan: Mahākappa 8/ Jhāna entry level: 2nd. *(T)*

Level 16 ***Appamāṇabhā*** – The heaven of Devas of unlimited splendor
Lifespan; Mahākappa 8/ Jhāna entry level: 2nd. *(T)*

Level 17 ***Parittābhā*** – The heaven of Devas of limited splendor
Lifespan; Mahākappa 8/Jhāna entry level: 2nd. *(T)*

Level 18 ***Mahābrahmā*** – Heaven of the great Brahmā
Lifespan – text illegible.
Jhāna entry level: 1st. (T)

Level 19 ***Brahmaparohita*** – Heaven of Deva priests & ministers of Mahabrahmā
Lifespan – text illegible.
Jhāna entry level: 1st. (T)

Level 20 ***Brahmapārisajja*** – Heaven of Devas in the retinue of the Mahabrahmā
Lifespan – text illegible.
Jhāna entry level: 1st. (T)

World Three – Which Includes 6 Heavenly Realms and 5 Other Realms (Kāmaloka)

The world of desire does not only include the remaining heavens, but all other levels of existence including the hells. The inhabitants of these six lower heavens are free of ills and have very long lives of enjoyment.

The heavens in this realm are magnificent and lavish.[125] There are no Jhāna entry requirements for these following heavens.

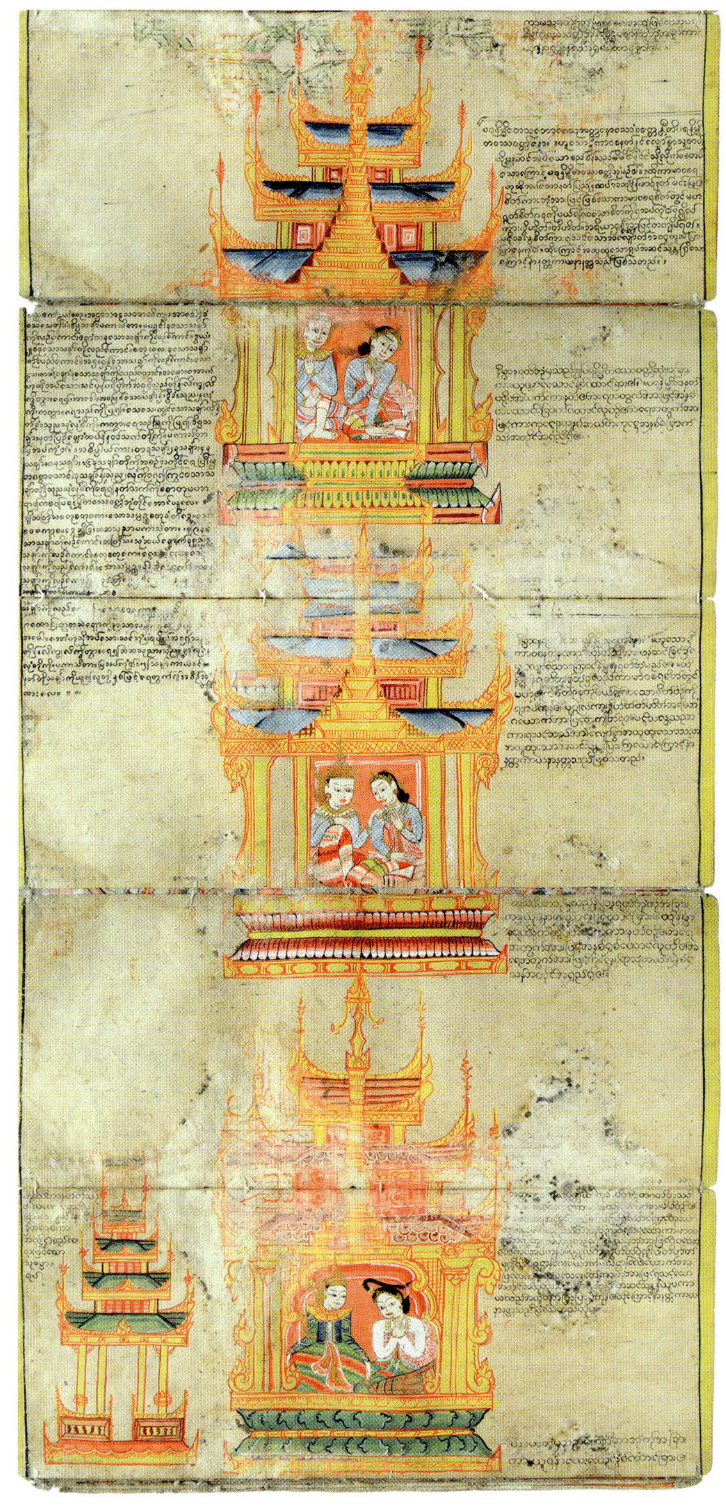

Level 21 ***Paranimmita-vasavattī*** – Heaven of Devas who find pleasure made by others (Manuscript pages 11 &12)

Lifespan and distance from nearest realm are not given.

This is also the home of the evil deity Māra[XXXII] who rules and who challenged the Buddha a number of times as he was on his way to enlightenment.[XXXIII] Māra is an evil deity exuding sensuous desires and wants, passions that overwhelm humans and stop the good from coming forth. Māra attacked the Lord Buddha in his quest for enlightenment for a period of six years before it was achieved. During that time his daughters tempted Buddha as he was meditating under the Bodhi tree and he also assailed him with thunderbolts. The defeat of Māra signified that Buddha had overcome his worldly cravings.

Level 22 ***Nimmānarati*** – Heaven of Devas who enjoy their own works (Manuscript pages 12 & 13)

Lifespan 16,000 Deity Years; equivalent to 9,216,000,000 human years. (T)

Level 23 ***Tusita*** – Heaven of blissful Devas and the residence of the Bodhisattvas (Manuscript page 13)

Lifespan 4,000 Deity Years; equivalent to 576,000,000 human years. Tusita Heaven is 42,000 yojana away from Yāmā Heaven. (T)

This heaven is the abode of all Buddhas who, for karmic reasons need to be born on the earth one more time and is also the seat of Maitreya.[126] It is considered the most beautiful of the celestial heavens. The open pavilion to the left of the main pavilion is a rest house named after Sudhammā, a wife of Indra (Sakka) in a previous birth.[127] The following text box is extracted from a Thai manuscript painting of Maitreya, the next Buddha, in Tusita Heaven.[128]

XXXII Māra is thought to be the absorption of the Hindu god Mavi, the king of demons, into Buddhism.

XXXIII In Burma, Māra is one of the Nats outside the pantheon of 37 Nat gods. He is called 'Man-nat' and is describes as a spirit of evil.

MAITREYA IN TUSITA HEAVEN

Maitreya, the future Buddha, resides in Tusita Heaven, awaiting and preparing for his last rebirth. Maitreya is destined to be born as the last Buddha of the present age. (A cult of Buddhist worshippers exists today praying that they will be reborn during Maitreya's birth and arrival into the world some 30,000 years hence.) This drawing extracted from a Thai manuscript painting shows Maitreya in rich apparel of a Tusita god, with his entourage of celestial attendants and musicians, floating through the skies of Tusita Heaven. Maitreya is totally ensconced in an aureole. Attendant figures wear elaborate costumes and small regal crowns. Some of them are musicians playing to announce the coming of Maitreya.

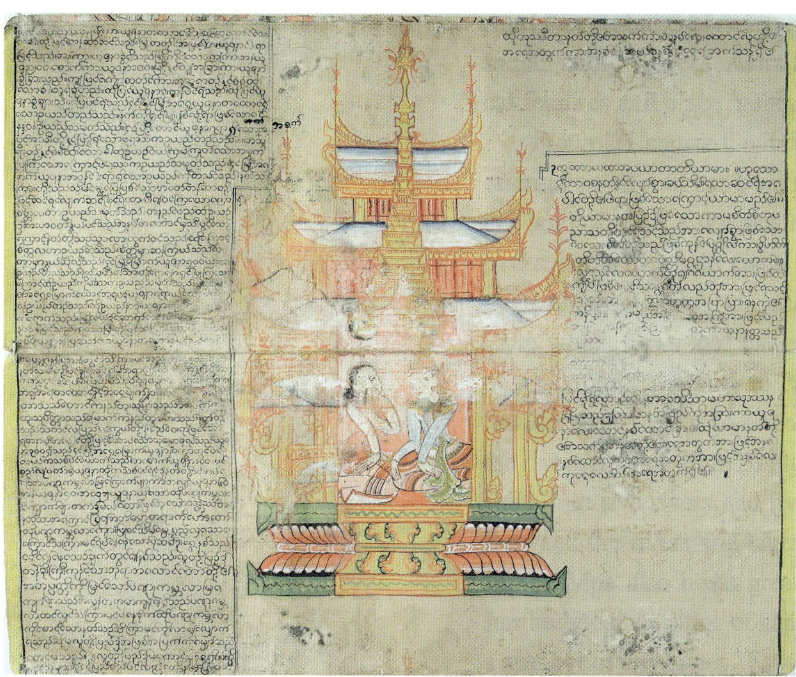

Level 24 *Yāmā* – Heaven of Yāmā Devas (Manuscript page 14)[XXXIV]

Lifespan 2,000 Deity Years; equivalent to 44,000,000 human years. (T)

In the heaven of the Yāmā Devas, King Suyāma reigns and is shown with one of his queens.

XXXIV The drawings of the Yāmā and Tāvatimsa heavens are both crowded by script. They have been damaged by water which caused ghost images that occured when the paper has been folded back while wet. The paper has cracked in places.

Level 25 *Tāvatimsa* – Home of Indra (Sakka) and the 32 devas (Manuscript page 15)
Tāvatimsa heaven is 42 yojana away from *Cātummahārājika* Heaven
The lifespan is 1,000 Deity Years equivalent to 36,000,000 human years. (T)

Tāvatimsa, at a low level among the heavens is, for most Buddhists the most interesting of the many heavens. Manuscripts of the cosmos have accorded special treatment to this heaven as the most detailed and elaborately drawn of all the heavens.[129] Tāvatimsa heaven plays an important role in Buddhist thoughts. It was visited by both the Buddha and Phra Malai. Tāvatimsa is a heavenly realm on the summit of Mount Meru, above that of the heaven of the four great kings who guard the cardinal quarters of the world. Mount Meru is the home of thirty-three gods, one of whom is Indra (Sakka), its ruler. Tāvatimsa is a heaven that can only be attained by the accumulation of merit. Phra Malai flies to this heaven to meet with Indra and to have discourse with Maitreya.[XXXV] This is the heaven where the Buddha passed his first rainy season after Enlightenment, during which he preached the Abhidhamma to his mother, Lady Māyādevi, who died shortly after his birth and was reborn into this heavenly realm. Traibhumikatha Tāvatimsa Heaven has the most abundant descriptions compared with the other heavenly realms.

XXXV The Phra Malai story is one of the core texts in Southeast Asia Theravāda Buddhism. (The others, which form the teaching and definition of the religion in Thailand, are the Trai Phum and the Jatakas, especially the Vessantara Jataka.)

Figure 23. Reproduction of the manuscript's drawing of Tāvatimsa Heaven[130]

The center of the scene is Sakka's palace, the Vejayanta, an elaborate structure with soaring, tiered roofs, enhanced by chedi-like structures on the roof and serpent-like finials.[131] Sakka sits in the center of the palace on the Pan-thu-Kab-balar couch surrounded by his four consorts: Sudhammā, Sujātā, Sujita and Sunandā. The consorts with elaborate hairdos are relatively simply dressed. Sakka is clothed in a kingly costume replete with crown. The costume, as befits Sakka, is green. The queen-consorts appear to be listening attentively.

Tāvatimsa Heaven normally also includes:[132]

1. The Culamani Chedi where a hair of the Buddha is enshrined. The hair was collected by Indra after Gautama's initial act of renunciation when he cut his hair to renounce worldliness. (In some cases the chedi is said to hold a tooth relic of the Buddha.)
2. The *sala*, or pavilion of Sudhammā, one of Indra's consorts, who was reborn in Tāvatimsa heaven. The gods come to the pavilion on the 8th day of every month to pay homage.
3. The Pinle Kathit Tree.
4. The Buddha, who visited Tāvatimsa Heaven to see his mother and preached the Abhidhamma to her, is depicted in 'the calling the earth to witness posture', or 'subduing Māra'.
5. Either the gateway to Tāvatimsa Heaven called the Cittakūta, or the Suvaññacitta gate which leads to a garden where on the full moon in May, all the great spirits of the earth, forest and skies gather together.

Tāvatimsa Heaven, as noted, was visited by the Buddha and Phra Malai. (See following text box. A depiction of Buddha's descent from Tāvatimsa Heaven may be found in this book's Prologue.)

PHRA MALAI IN TĀVATIMSA HEAVEN[133]

Phra Malai is here depicted on a visit to Tāvatimsa Heaven meeting Indra, the ruler of the Heaven, at the foot of the Culamani Chedi. The chedi, a reliquary, contains a hair of the Buddha.[134]

Phra Malai, a saintly monk, through meditation gained the powers to move between heaven and hell and to travel through various cosmic realms.

A discourse between Maitreya and Phra Malai ensued, this covered: Maitreya's merit-making which emulated that of the Buddha. It also established conditions in the human realm and its future, and; finally predicted the deterioration of Buddhism and the degradation of human nature. The discussion posited after the teachings of the Buddha had been on earth for five thousand years, human life would have deteriorated so that it involved increasing amounts of incest, promiscuity, violence and war, with the destruction of most of human life. However, a few wise people would retreat into forests and would emerge to form a new society after which Maitreya would be born. Maitreya told Phra Malai that members of the human realm should observe and keep the precepts and listen to and observe the Vessantara story if they wished to be reborn when he became the Buddha. He stated by doing so they would attain Nibbana.

Level 26 *Cātummahārājika* – Heaven of the four kings of the cardinal quarters
The lifespan is 1,000 Deity Years equivalent to 36,000,000 human years. (T)

The *Cātummahārājika* heaven is the lowest of the heavens; it is located on the great column of Mount Meru. Four lords monitor and rule the four cardinal corners of the world,[135] from Mount Meru to the iron walls of the Cajravada mountain range. The Lords who live in *Cātummahārājika* Heaven enjoy large gem castles. All of these kings serve Indra, Lord of Tāvatimsa Heaven. Three days of every lunar month the four kings either go or send observers to view the status of righteous and virtue in the world of men. These findings are reported to the assembly of the Devas of Tāvatimsa Heaven. On the command of Sakka, the four kings and their followers, the legions of the supernatural, stand guard to protect Tāvatimsa Heaven from attacks by the Asuras which now and then threaten to destroy the realms of the Devas. The kings have also pledged to protect the Dharma and the followers of the Buddha (Manuscript pages 16 & 17).

These four kings have enjoyed popularity in China since the fourth century. The Tang Dynasty (618–907 C.E.) instituted their veneration. Monasteries have halls devoted to them in China and elsewhere in Asia. Each king is thought to have 91 sons as well as eight generals and other followers who help guard the king's quarter of the world.

This manuscript is unusual in having all four kings illustrated; often a single kingly figure represents the kings and *Cātummahārājika* Heaven.

Lord Virūdhaka rules all the devas on the southern side. Lord Virūdhaka is king of the yaksha. He battles against darkness to protect the base of goodness in people.

Lord Vessavaṇa rules all the devatā on the northern side Lord Vessavaṇa is also king of the yaksha in his domain.

Lord Dhataraṭṭha rules all the devatā on the eastern side. This king is a lute player, purifying the thoughts of men.

Lord Virūpākkha is not depicted, however his symbols are. These are the Nāgas and Garudas in his domain and whom he controls. Lord Virūpākkha rules all the devatā on the western side

GARUDA AND NAGA

The garuda, half bird-half human, is the king of the mystical birds and the arch enemy of serpents. The garuda serves as the steed for Vishnu, a chief deity of Hinduism. The Naga is depicted as a semi Devine being with, sometimes, a human head. The antagonism arises from a falling out of the garuda's mother with Kadru, the mother of snakes. The two are separated in Lord Virūpākkha's chamber by a golden crown. (Source: Fischer-Schreiber, Ehrhard, Friedrichs & Diener 1999)

The Lowest Five Levels of Existence

Below the heavens five lower levels of existence remain to be addressed. This manuscript contains three of these. Burmese manuscripts usually address Level 28 of the Asuras but seldom Level 29, the level of Preta. All manuscripts show some of the various hells of Level 31. The hells of *Niraya* are not the only hells; rebirth in *Asura* or *Preta* would still be most unfortunate. The hells of *Niraya* are much more segmented and structured than those of *Asura* or *Preta*; all depending on one's karmic circumstances would lead to an awful existence.

The Realm of the Asura (Level 28) (Manuscript pages 17 & 18)

Beneath Mount Meru, one of the kings of the Asura hell looks directly at the viewer. He is flanked by two queens in diaphanous gowns. They are ensconced in a small regal pavilion, the spire of which penetrates the lower reaches of the sacred mountain, and seemingly reaches with futility towards the realms of the 26 heavens above.

The Asura realm is at the foot of Mount Meru and much of it is in the seas and the Great Ocean. The Asura, once demigods, as powerful as the deities in *Cātummahārājika* and *Tāvatimsa* heavens were deposed from *Tāvatimsa* Heaven for acts of jealousy, and drunkenness. They constantly fight to regain their lost kingdom on the top of Mount Meru, but are stopped by the four guardian kings who reside in the palaces of the *Cātummahārājika* heaven on the terraces of Mount Meru.

The Asura are formed into many leaderless warring groups.[XXXVI]

XXXVI Burmese legends describe three types of Asura, namely (1)Wei-pa-seik-ti Asura (2) Wi-ni-pati-ka Asura and (3) Ka-la-kein-si-ka Asura. The following narrative, in nature, is a folk tale whose source is Sai Keing Hserng, Burmese author and translator.

(1) **Wei-pa-seik-ti Asura** – They once resided in **Tāvatimsa** Heaven on the very summit of Mount Meru with the other deities. where they love to drink and were drunk most of their time. A man Mar-ga and his 30 friends who had much good merit by the construction of bridges, water reservoirs, and other works during their life as human beings and, they became deities in the **Tāvatimsa** Heaven as a result after they died. Later, on seeing the drunken Asura, Mar-ga, and others ordered by Sakka, threw the whole lot off the summit down to the base of Mount Meru. The Asura in their drunken state could not resist. The realm of Asura was very much like **Tāvatimsa**, except in three things; (a.) The Thu-dha-mar Zayat, (b.) Wei-Za-yarn-dar Pyat-that and (c.) the Ka-thit tree. While Asura lack the first two they did have a Thakhut tree instead of the Ka-thit tree. The Asura were too drunk to realize that they are not in **Tāvatimsa**. Once a year, at the beginning of the rainy season, instead of scarlet red Ka-thit tree flowers, they found a dull yellow or white Thakhut flowers. When they realized their situation, they march for a battle to return to their former domain. At this time of the year the sound of thunder and lightning are said to be the sound of war drums in the battle between Indra from **Tāvatimsa**, and the Asura. The Asura were once powerful and as divineas the deities in **Cātummahārājika** and **Tāvatimsa** heavens. They were also once part of the seven sensual realms.

(2) **Wi-ni-par-ti-ka Asura** – They were also divine Asura, but not as powerful as those from **Cātummahārājika** and **Tāvatimsa**, and other heavenly realms. They have rare dresses and ornaments, and special food to eat. Some called them ' little divine', because they are less powerful, and have to relied on the **Bom-ma-soe** deities, and those who have their own residence.

(3) **Kar-la-kein-si-ka Asura** – They are actually Byeik-ta (Preta) who are included in the realms of the Asura. They lived on the bank of the oceans, the Ganga and other rivers. Even during the time of two and three Buddhas who are graced with enlightenment, they still cannot eat, or have nothing to eat, and are in a state of severe hunger all the time.

The Realm of the Preta (Level 29) (Manuscript pages 21 & 22)

The Realm of the Preta (Level 29) is introduced following the two pages of discussion on the aspect of the Cakravāla hells. Is this an anomaly? Patricia Herbert noted that 'Burmese manuscripts omit the realm of Peta'.[136] The text and drawing in the manuscript indicate that this is not the case for this manuscript. The following manuscript page which also contains a mountain and two trees which have been identified by a Burmese scholar and are as annotated.[XXXVII]

Descriptions of this level of existence from a number of sources are long and fearsome. The following narration is deliberately brief.[137]

The inhabitants of this level, often termed 'Hungry Ghosts' are born here after a former life of wicked deeds, primarily great envy of others and dismissal of the dharma and the Four Noble Truths. As the result of this bad Karma, these inhabitants have overwhelming needs for particular foods, items or other unattainable objects. Some are reborn at this level for their stinginess, for example if they had ample funds and never gave alms; others because they mixed bad rice with good to make more profit. They cheated monks out of temple land.

Many descriptions of these Ghosts and the torments of their lives exist in this realm. There are many types of hungry ghosts and many conditions in which they exist. Some suffering ghosts reside in palaces of gems surrounded by gem walls and are dressed like devas; others have elephants and slaves at their beck and call; others can only lie on their backs hoping for food. The food of the latter is horrible and unthinkable, including feces, urine, pus and the like.

Hungry Ghosts take many forms; ones shown here has a great bloated body and a miniscule head; another has a great head and minuscule body. Others with more human aspects are shown impaled on a tree or crouching around it. In other manuscripts, suffering ghosts were often portrayed by the artists as just skin and bones while the plight of others are amplified by animal heads and enlarged testicles.

Preta is for those without morality, who do not follow the Dhamma, who disturb others and who trouble the monks. They are consigned to an awful state of life with abnormal size heads and small bodies. They are very hungry and cry for food and alleviate their hunger by eating feces and drinking saliva. They await the arrival of the next Buddha to relieve their situation. (T)

For those who disrespect the triple gems and insult the ten precepts, who steal, who are wicked to benefactors, and who are persons who say 'bad is good and good is bad' will suffer in minor hells even though they are free of the major ones. They cannot stand by themselves. (T)

XXXVII An 'Adjhan Wat' (Temple Professor) of a temple in Chiang Rai Province, Thailand; formerly, Abbott of a Burmese Temple with 25 service in the Sangha.

Figure 24. Painting of the inhabitants of Preta,[138] *Northeast Thai / Cambodian Manuscript painting*

This manuscript drawing is either from Thailand's northeast region, or from Cambodia. The manuscript's script is Khom, an ancient form, of Khmer that uses Pali as its language. Khom is a sacred language used exclusively for prayers, sacred texts and mantras. The drawing while primitive is quite expressive and fulfills its didactic requirements well.

The drawing depicts the classical categories of miserable spirits that inhabit the Preta realm. It is similar to illustrations to be found in the Traiphum: the story of the three worlds of existence.

Phra Malai flies above the sufferers in Preta. He witnesses their punishments and gives them comfort. He also flew further into hell to comfort the sufferers there. All begged him to intercede on their behalf and to relieve them from their situation; this is denoted by the figure with raised arms and hands. Those in Preta and hell asked him to have their relatives make merit and transfer it to them so they could be reborn in a better realm; possibly the human world where they could make merit on their own behalf.

The painting shows Phra Malai in an elaborate robe, floating against a monotone brown background. A modest aureole, fringed in gold, surrounds his head. The emaciated figures in the lower half of the painting are beset by hunger and the fires of hell. One female figure has a goat's head.

Figure 25. Phra Malai Visiting Preta,[139] *Thai Manuscript Painting*

Niraya: The Hells of the Cosmos (Level 31)

The 'hells' of Buddhist cosmologies are, for the artist and, perhaps, for the viewer one of the more interesting sections of the manuscript. The hells, like most of the parts of the cosmos, are compartmentalized. There are major hells, minor or auxiliary hells and the Lokanta hells. The major hells depicted in this manuscript are only 'hot 'hells; 'cold hells' are not shown or discussed.

The following was noted by John Snelling:

> The torments of the Buddhist hells equal if not exceed anything conceived by Hieronymous Boch and other Western infernologists. The human imagination seems to become unusually inspired when it comes to devising extreme forms of horror and torment- and the early Buddhists were no exception.[140]

As will be demonstrated, the artist is not constrained by set boundaries, and thus imaginations have been able to roam freely. Hell has important places both in the thoughts and the art of Theravāda Buddhism. Hell is discussed and explored in the Phra Malai stories and the Jatakas (the stories of the 547 previous lives of the Buddha). Paintings in the manuscripts of Phra Malai take great interest in the hot hells. Phra Malai flew to the hells to view the suffering of the inhabitants. Indeed, a sizable part of each Phra Malai manuscript is devoted to the sufferings of those incarcerated in the hells. The main purpose of the drawings of the torments of hell is to show the results of bad karma.

Niraya: Translation of Manuscript Page 19

❊ ❊ ❊

Enjoying pleasure by day and suffering by night was the **Asura.** The hell have 8 levels, from top to bottom were, **Samjiva, Kalasutra, Samghata, Raurava, Maha Raurava, Tapa, Maha Tapa** and **Avici** The floor, wall and ceiling of each hell was made up of hot iron.

the (indecipherable…) Flame and smoke flying up as high as a hundred yojana. the (indecipherable…) The sufferers in these eight levels of hell were like different sea creatures residing at different depths and levels.

Each level of hell have four main gates opening on each side. And each gate have eight lesser hells, namely, Hot-iron, Fire-water (liquid flame?), Hot ash, thorny Lat Pan (*Bombax ceiba*), boiling copper, thorny cane, Bin-poke (waste & rotten offal). All together there are 32 lesser hells on each level. In total there are 64 major and lesser hells.

After suffering their terms and being discharge from the major hell, they had to suffer their terms in the lesser hells, the duration was decided by the hell wardens according to their sin, as a jail warden decides the fate of a convict in the human world. The hell wardens were *Kar-lu-pa-kaw, Yar-ma-law, Yar-ma-kaw, Yar-ma-taw, Thi-ri-thu-htaw-sa, Yar-mar-ti-ka and the most supreme, Yar-ma-ya-zar. Wa-tha-wun-na* the king of the **Preta,** King Thor-ma, King Yar-ma, Wa-tha-wun-na, the king of Belu[XXXVIII] and other Hell wardens did not torment the sufferer, only the evil deeds and sin they have committed were tormenting them. The number of sufferers in hell was uncountable.

XXXVIII The various Asura kings are also described. Among these is the giant king Rāhu (Burmese WathaWanna) who stands 4,800 yojanas tall and has a head circumference of 900 yojanas. A Preta king is also mentioned (Burmese Wat ThaWana) as well as the Yama king of the guardians of hell (Burmese Thaw Ma).

The **Avici** Hell has flames 100 yojana high and the whole cauldron is 37 yojana in width. With the inclusion of the *U-that-da-ret* or lesser hells the **Avici** Hell have a total area of 10,000 yojana and there is no place like it. The sufferers are packed like compressed mustard seeds inside a container.

In every hell, there is no one walking, sitting, standing or laying down. Of the six sensual minds to ignore, they have only the sense to realize and have no sense to ignore it.

In the 8 level of hells, each have four main gates opening on all four sides, with 8 lesser hell in each.

The so-called forbidden ones, those who have suffered in hell came back to live with one mind that was the animal instinct. Even in their embryonic stage, they developed with *ar-ku-tho* (evil mind) and once they come to life, they have 12 *Ar-ku-tho* (evil instinct), eight great *Ku-tho* (good instinct) and 7 animal instinct, which made up the 37 levels of conscious mind of a living being.

✲ ✲ ✲

The Major Hot Hells

The foregoing translation conforms to the *Abhidharmakośabhāṣyam of Vasubandhu*, and of King Lithai's work of eight major hot hells and the accepted version which describes them as layered one on top of the other. The Burmese manuscript here describes the floors, walls and ceilings of being made of hot iron and the eight levels of hell were at different depths. A summary of the eight great hot hells is shown in the following table and Figure 26. The figure shows these hot hells under the continent of Jumbudvīpa. In the *Abhidharmakośabhāṣyam*, and *Traibhumikatha*, and the foregoing translation, Avici is described as a huge area, the last and lowest hell; and the other hells smaller.

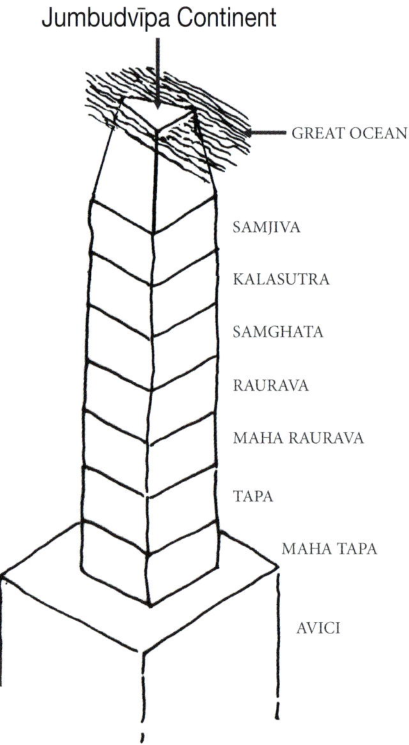

Figure 26. The Eight Major Hot Hells under the Continent of Jumbudvīpa

1. Samjiva – The hell with Death & Life Cycle.
2. Kalasutra – The hell with the Black Wire torment.
3. Samghata – The hell with the Stone Slab torture.
4. Raurava – The hell of Groans and Moans.
5. Maha Raurava – The hell of Great Groans and Moans.
6. Tapa – The hell of Fiercely Burning Fire.
7. Maha Tapa – The hell of Great, Fiercely Burning Fire.
8. Avici – The hell without Interruption.

In the manuscript with the exception of Hot Hell 1, the major 'hot' hells contain boxes with great flames emanating from them; the faces of the sufferers encased inside, looking out. Descriptions of minor hells in the main references exceed those of the major ones. The descriptions of the hot hells in the *Abhidharmakośabhāṣyam* are somewhat sparse, as are those in King Ruang's *Trai Phum* as well as King Lithai's *Traibhumikatha,* which is parallel to one another, as each is based on the same writings.

In the following translations of portions of the manuscript the reasons for incarceration in the various hells are simple and direct with no room for ambiguities. This manuscript is a teaching tool and is designed to perform that function well. The messages in the manuscript seem to be directed at a young audience with references for needed respect for parents, teachers and monks. Also included are repeated references to the precepts. The violation of a precept finds mention in a number of the hot hells as reasons for being there. There are many types of precepts in Buddhism; five is one number; eight and ten are others. The precepts are not canon law, but are injunctions for rational behavior and to promote human wellbeing.[141]

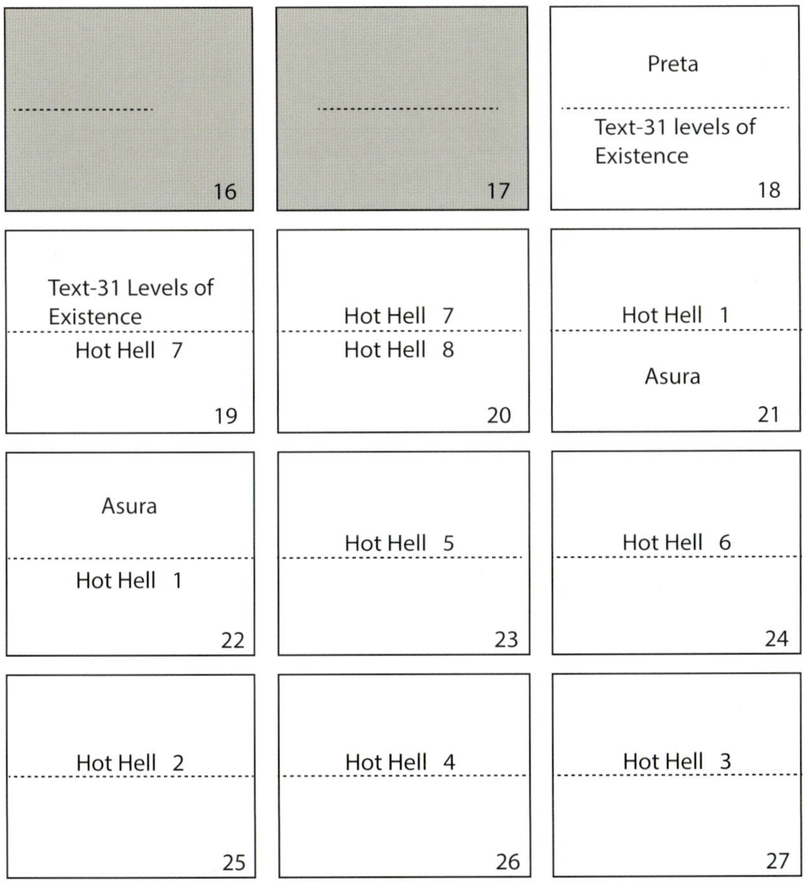

16	17	Preta / Text-31 levels of Existence 18
Text-31 Levels of Existence / Hot Hell 7 19	Hot Hell 7 / Hot Hell 8 20	Hot Hell 1 / Asura 21
Asura / Hot Hell 1 22	Hot Hell 5 23	Hot Hell 6 24
Hot Hell 2 25	Hot Hell 4 26	Hot Hell 3 27

The location of the hells in the manuscript wanders and is not orderly in the sense that it lacks serial precedence. The sketch from manuscript pages 16 to 27 shows the location in the manuscript of the 'hot hells'. Although best efforts have been made; it has to be noted that either the interpretations, or the manuscript drawings themselves may be wrong. It should be noted that the manuscript drawings of the hot hells stray from the traditional descriptions of the types of hell.

In the drawings of the hells a series of burly men are encountered (see following text box). These are the servants of Yama. The translation of the manuscript states that hell wardens did not torment the sufferer, the evil deeds and sins they have committed were tormenting them; however, as that may be the manuscript drawings show otherwise.

YAMA: [142] THE RULER OF HELL

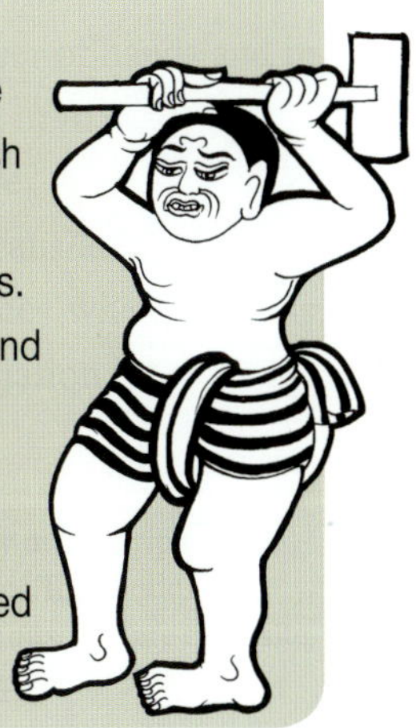

The beefy servants of Yama are shown working hard in the hot hells, but who was Yama? Yama was once a king who during a bloody war wished himself to become the ruler of hell. In keeping with his wish he was reborn as Yamamin (Burmese: min = king) along with his eight generals and 80,000 servants who accompanied him to his kingdoms in hell. In hell he is in charge of the karmic process. As dead souls are reviewed, he reminds them that they alone are responsible for being in hell and for their punishment. The servants of the Yamamin, according to Phya Lithai, were relegated only to the minor or auxiliary hells.[143] In the manuscript we find them on active duty in all of the manuscript's hot hells. Some authors explained that this duty of the Yama's servants is temporary for 15 days, after which they suffer the tortures of this hell and are replaced by another group of servants of the Yamamin, and so on, in rotation for a long time.

Hot Hell Number 1 – *Samjiva*: The Hell of the Cycle of Death and Life, or Constant Repetitions (Manuscript pages 21 & 22)

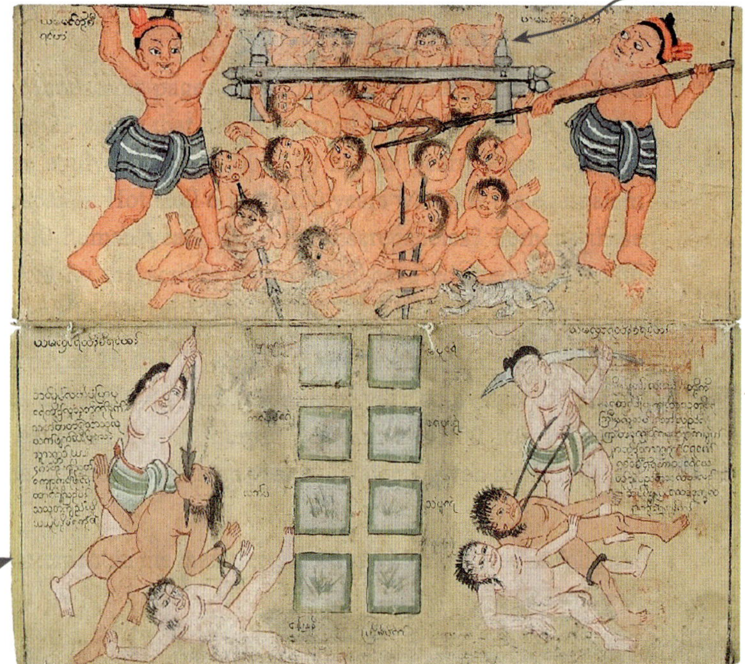

A servant of Yama doing his work. (T)

The minor hells here are for those who do not control, or do not observe polite or normal bodily or verbal and mental actions, or have mean thoughts. These people suffer here, not in the major hells. (T)

These minor hells are for those who kill animals. (T)

In Samjiva, the body is first crushed and reduced to dust. Then a cold wind blows and revives the bodies and restores feeling; however, again the crushing begins.[144] An alternate description is one of having the inhabitant's dismembered with various kinds of sharp blades after which the limbs and body parts are scattered. Then a wind blows and the scattered remains reassemble, and the cutting and dismemberment begins again.

Hot Hell Number 2 – *Kalasutra*: The Hell of the Black Wire (Manuscript page 25)

Beings assigned to this hell are placed on a floor with red-hot iron wires which cut their bodies into parts. Alternately, the bodies of the sinners are wrapped with black iron cords and these are used as guides to saw the bodies into pieces. Note that the minor hells precede the major hells as illustrated in this manuscript.

The picture has a typical hot hell cauldron and then eight minor hells leading to a partial hot hell one. There are also four guard boxes instead of the usual two. However, the illustration on the reverse side of heavens is in order. The conclusion is drawn that the picture, one of the products of many hands that worked on the manuscript, may be a mistake which could not be rectified with white-over and repainting.

Minor Hells are for those who torture other beings and who are wicked and forget to do good. (T)

Minor Hells are for those who do not respect, or who mistreat their parents. (T)

The lifespan of the sufferers is 1,000 years or 32,000,000 human years. The guardians of this hell are called 'Upaka'. (T)

Hot Hell Number 3 – *Samghata*: The Hell of Stone Slabs (Manuscript page 27)

In Samghata huge rocks crush the sufferers to a bloody jelly. When the rocks move apart, life is restored and the process starts again. Alternately, the rocks are molten lava.

The lifespan of the sufferers is 4,000 years or 576,000,000 human years which is equivalent to a day and night in this hell. The guardians of this hell are called 'Yamataw'. (T)

These minor hells are reserved for those who use bad language to their parents and mistreat teachers and who committed the ten kinds of wrong actions including breaking the precepts which guide monks. (T)

These minor hells are reserved for those who lack charity and those who spoil or pollute, or spit into the well, or lake. (T)

Hot Hell Number 4 – *Raurava*: The Hell of Lamentation (Manuscript page 26)

Victims are constantly tormented in this hell, with little hope of escaping the flames. They run around looking for refuge from the burning coals underfoot and when they think they have found shelter they are locked inside it, and it blazes inside while they scream.

The lifespan of the sufferers is 2,000 years or 44,000,000 human years. The guardians of this hell are called 'Kala'. (T)

These minor hells are for those who have five kinds of jealousies and who have no respect for elders and those who believe in devils and forego the knowledge of the triple gems. (T)

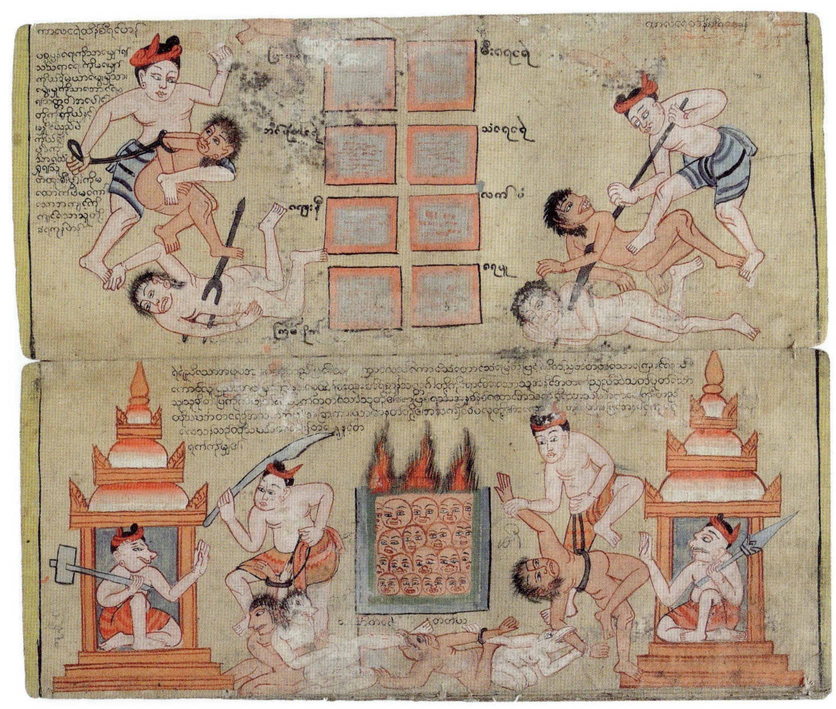

Hot Hell Number 5 – *Maha Raurava*: The Hell of Great Groans and Moans (Manuscript page 23)

The torments are similar to Hell 4, but much more intense. In this hell, devil animals torment sinners and eat their flesh.

The lifespan of the sufferers is 8,000 years or 32,000,000 human years. (T)

Those who do not listen to their parents, teachers, or monks, or those who do not follow the five precepts, or who steal from the temple or who kill animals or people are consigned to his hell. (T)

Minor hells are reserved for those who do not listen to their parents, teachers, or monks. (T)

The minor hells are for those who do not respect, and who use bad language to their elders. (T)

Hot Hell Number 6 – *Tapa*: The Hell of Scorching Heat (Manuscript page 24)

The inhabitants of this hell are pierced by hot iron stakes or spears. A great, excessive fire, awaits the sufferers while piercing continues until flames come out of the sufferers' nostrils and mouths.

The lifespan of the suffers is 16,000 years, or 16,000,000 human years. (T)

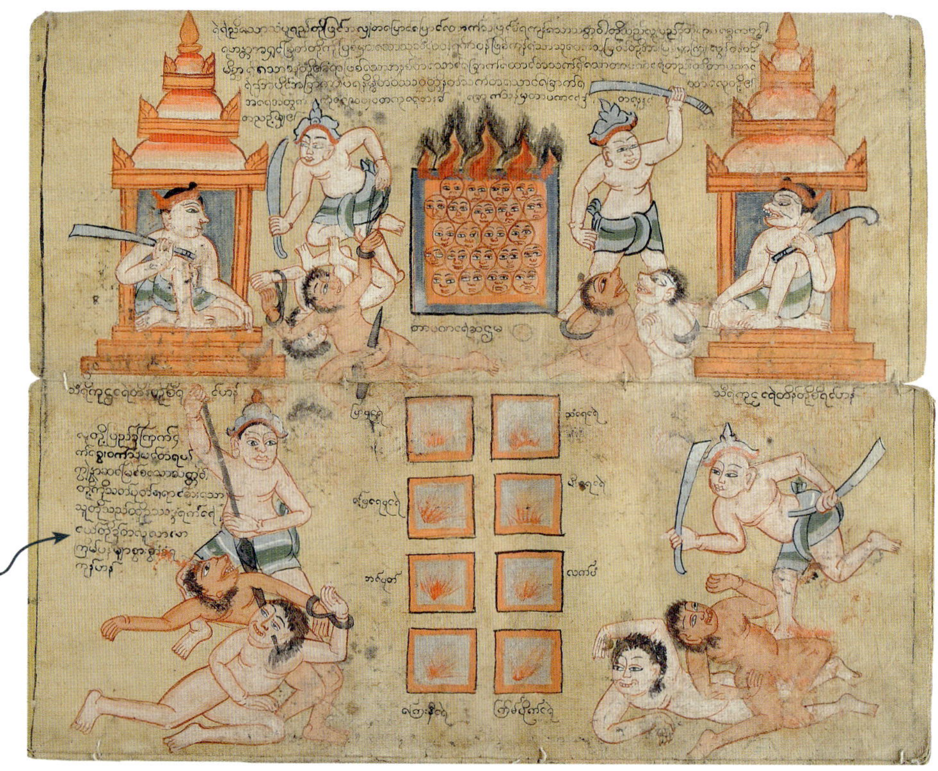

Tapa minor hells are for those who kill animals. (T)

Hot Hell Number 7 – *Maha Tapa*: The Hell of Fiercely Scorching Heat (Manuscript pages 19 & 20)

Maha Tapa is the great heating hell with tortures similar to that of Hell 6, *Tapa*. The difference is that the instruments of torture vary and include tridents (not shown).

Poured on with molten iron to death, but revived to repeat the circle, again and again. Those who were greedy; who had robbed, cheated or steal from others, those who have killed in anger and hate will suffer in this Thi-ri hell for the duration of 500 (indecipherable…) A lifespan equivalent to 500 years in human world in Cātummahārājika Heaven (indecipherable…) and (indecipherable…) (T)

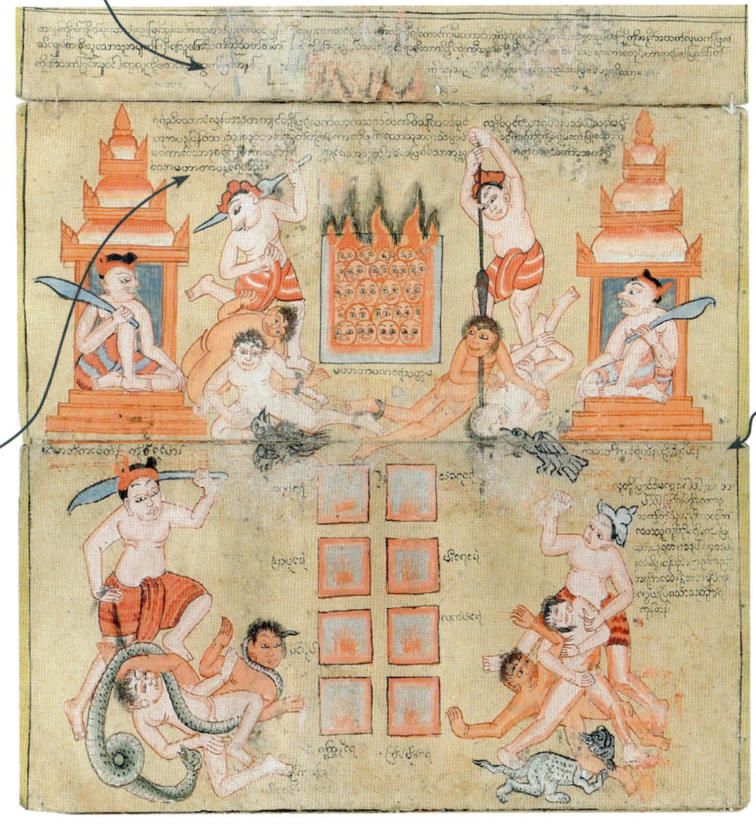

The minor hells are for those who show disrespect to their elders and never listen to those who teach and guide them. (T)

Impaled from the right to left flank with red hot spears or stakes, are for those who do not believe in the three treasures, those who do not heed the words of the sacred ones, those who do not respect their parents, those who always thought and practiced evils, will have to suffer in this Maha-otaoana hell with a lifespan of half the anantara kappa. (T)

Hot Hell Number 8 – *Avici*: The Hell without Interruption (Manuscript page 20)

This hell is for those who have killed parents, monks and teachers, or who destroy chedis. The fires of this hell reach 100 yojanas high and are 100 yojana deep. (T)

Avici, the lowest and largest of all hells, is without respite; suffering in the other hells is interrupted from time to time, but not in *Avici*. This is the hell where Devadatta,[XXXIX] has been incarcerated. Others in the same position are the doers of the principal deeds of great evil. They remain in this hell for an Arnantara (infinite) kalpa. As they reach the three quarter mark of their term a great fire consumes the kalpa; thosewho have not consumed with time left on their kalpa are swept to a similar hell in another universe. They cannot ever escape the consequences of their evil deeds.[145]

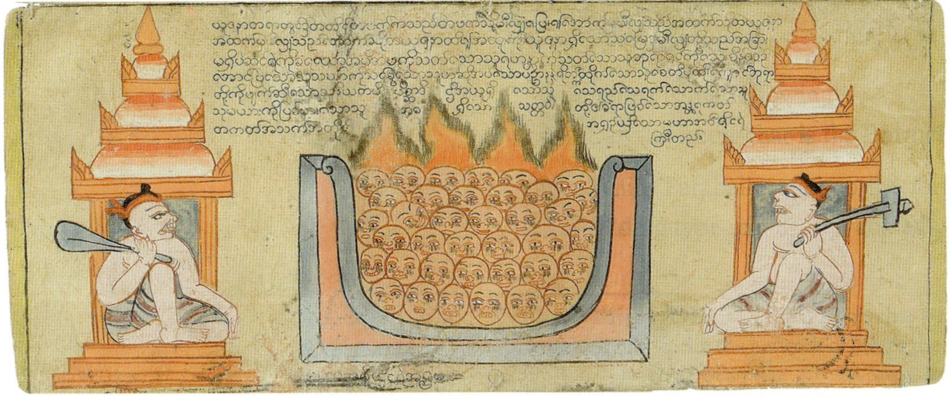

Avici's cauldron is the largest of all the hot hells, reflecting its size as the largest of all the hot hells. There are no minor hells illustrated in the manuscript.

XXXIX Devadatta, a cousin of the Buddha, who became a highly respected member of the Sangha. Eight years before the death of the Buddha, he schemed to become head of the Buddhist community and to murder the Buddha. He then instigated a schism among the monks.

Other Hells

Aside from the major 'hot' hells, the manuscript has two other types of hells, the minor or auxiliary hells and the Lokanta hells which are located beyond the Cakravāla World.

The Minor, Utsada, or Auxiliary Hells

Part One noted that there are minor hells- four at each of the four gates of the eight great hot hells, for a total of 128 minor or auxiliary hells. This Burmese manuscript identifies eight minor hells.[146] The hot hells in this manuscript each have only eight minor hells, for a total of 64 minor hells. Since these eight hells are the same for all the great hells, there are only eight types of minor hells.[XL]

The minor hells in the manuscript are shown as orange, or green fringed squares with white fill. The minor hells are associated with each of the major hells. The inhabitants of these minor hells are there for specific reasons which vary with the major hells they are condemned to.

TABLE 5. MINOR HELLS: Their Names and Torments

Left Side (top to bottom down)		Figure extracted from Hell 6 – *Tapa*: The Hell of Scorching Heat	Right Side (top to bottom down)	
Burmese Name	Torment		Burmese Name	Torment
Pyar Pu	Hot ash or chaff		Than	Heat Iron
Ye Pu Hell	Hot Water or burning river		Mee	Fire
Bin Pote	Human waste & decaying corpses		Let Pan	?
Kye Niana	Razor blade road		Kyein Pite	Canes full of thorns

XL The Burmese Wei-da-du-ta scripts-state that there were 5 lesser hells on each gate of the four major hell, altogether 20 lesser hells in each level. The Reverend Manhle Sayadaw described in his Ma-ga-De-va poem mentioned 10 lesser hells.

The Lokanta Hells

Hells external to the Cakravālas may be found sequestered between other world systems. Another hell called the Lokanta hell can be described as follows.

> There are three cakkavāḷa close to one another like three ox carts. They are placed side by side like three overturned begging bowls placed next to one another. In each case between these three cakkavāḷa, there is a Lokanta hell.[147]

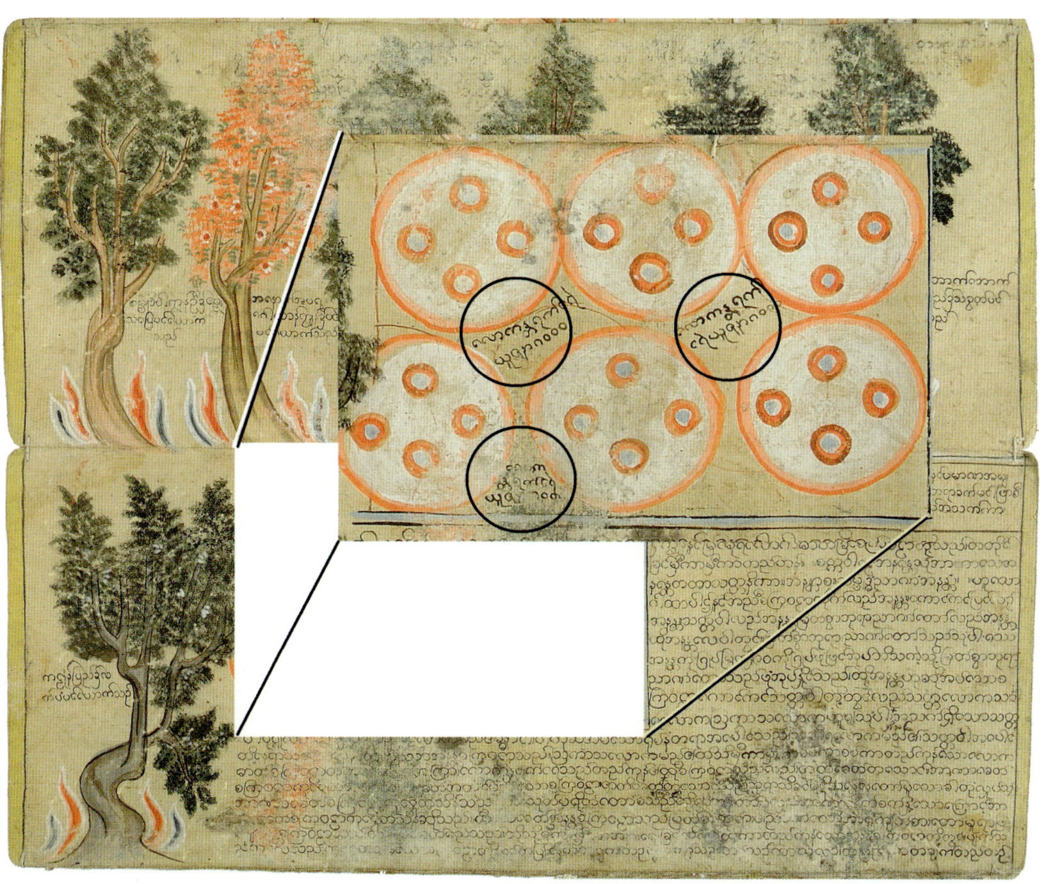

Figure 27. The Lokanta Hells (Manuscript page 32)

The manuscript shows three Lokanta Hells. They are shown sandwiched between six Cakravālas in the drawing of trees (see Figure 42, from which Figure 27 is extracted). The six planets indicated by the orange-bordered white circles have four continents of the world system within each of them. These continents are shown by the same orange circles; the three Lokanta hells may be found between the world systems. Lokanta Hells are 8,000 yojanas wide and deeper than the gold sand layer. The bases of the Lokanta Hells rest far below in the water layer that supports the cosmos.[148] The Lokanta Hells are reserved for those who have harmed parents, monks or Brahmin teachers who have not kept the precepts, and those who have caused monks to quarrel. These sinners are born in the Lokanta Hells. The Lokanta Hells are dark, awful places without divine abodes and populated with inhabitants who are 6,000 *wa*, or three kilometers tall. The inhabitants are subjected to the stress of constant death and rebirth. Death is caused by the search for food which leads them to attach themselves to the walls of the adjacent Cakravālas. Because the hell is dark when they become hungry they fly to find food. They discover another inhabitant, but not knowing it is a fellow sinner they begin to consume each other. Holding on to each other they loose their grips on the Cakravāla wall and fall into the water below and disintegrate not unlike 'feces that fall into the water'.[149] They die and are reborn again to suffer the same fate for great lengths of time. The following, Figure 28 shows a Thai version of Lokanta Hell. Clearly seen are two Cakravālas worlds and the seven Sattaparibhanda Mountains with the inhabitants clinging onto the iron walls of the Cakravāḍa mountain range and moving through the water seeking sustenance.

Figure 28. A Lokanta Hell from a Thai Manuscript[150]

PART 4

Cosmography

The question arises: Is this cosmography consistent with the teachings of the Buddha? Henry Alabaster, in his 1871 book 'The Wheel of the Law' answers succinctly,

> Had the Buddha taught cosmography as it is in the 'Traiphoom' he would not have been omniscience, but by refraining from a subject which men of science were certainly eventually to ascertain truth of, he showed his omniscience.[151]

The Cakravāla's Cosmography: Translation of Manuscript Page 18

It will be seen that the following translation from the manuscript parallels Part 1. As noted in Part 2, for commonality with the other parts of the book, the Burmese nomenclature used for realms, places and beings has been rendered into Pali (and are shown in **bold**). The transliterated Burmese names may be found in Appendices A and B.

☼ ☼ ☼

At the base of Mount **Meru** there are three small mountains which form a tripod on which rests Mount **Meru.**[XLI] In between these mountains is a plain of gem sands. There, a space of 10,000 yojanas wide is to be found the land of **Asura**. It was counted as part of **Tāvatimsa** Heaven, and was one among the four ten-thousand. The four ten-thousand yojana were, the Śakra city of **Tāvatimsa** Heaven, The south island, **Jumbudvīpa**, the Avici Hell and the land of Asura, in which each of them have the area of 10,000 yojana.

All the above realms were just part of the universe which have a length of 1,234,000 yojana and a width of 691,350 yojana. The earth was made up of two layers, each was 12,000 yojana thick, *Pan-thu-pa-hta-wi* (soil), which sits on top of the *thi-lar-pa-hta-wi* (stone) layer. All this earth was afloat and on top of a 48,000 yojana deep body of water. These masses of soil, stone and water are suspended above an air layer 96,000 yojana deep. Instead of width and length, the earth was usually described only in thickness of 24,000 yojana surrounded by oceans of 84,000 yojana deep, which were part of the universe.

At the center on the universe is Mount **Meru** with 86,000 yojana[XLII] submerged in the ocean and 86,000 yojana which towers above the water. Spiraling up from the base of Mount **Meru** are five levels of residence terraces four of which are occupied by **Virūpākkha, Dhataraṭṭha, Vessavaṇa** and **Virūdhaka** the Kings of the cardinal directions all ofwhich are 4,200 yojana away from human world. The sun, moon, stars, constellation and other celestial bodies, orbit around the middle height of Mount **Meru**.

Mount **Meru** is surrounded by seven ranges of mountains and seven seas. The inner most range was the highest and reduces by half for the adjacent one, and so do the depths of the seas. The seven ranges were, 1. **Yugandhara** Mountain Range, 2. **Īsadhara** Mountain Range, 3. **Karavika** Mountain Range, 4. **Sudassana** Mountain Range, 5. **Meemindhara** Mountain Range, 6. **Vinataka** Mountain Range, 7. **Assakaṇṇa** Mountain Range. As the inner range was higher than its outer one, most scholars claimed it was the *Pyin-tein*, or outer shallow. But some scholars, [XLIII] based on the south island, were objects have no shade at noon time, claimed that it was *Ar-twin-tein* or inner shallow. Which means, the height of the range and its adjoining sea have the same height and depth.

From the outermost **Assakaṇṇa** mountain range to the wall of the universe, the whole area is occupied by a salt-water ocean.

Mount **Meru** has four colors: Silver on the eastern side, Emerald on the south, Crystal on the west and Gold on the north side, which reflects the color of the ocean. Thus, *Si-ra-thar-ga-ra* the milk-like silver white color ocean on the east, *Ni-lar-thar-ga-ra* emerald green ocean in the south, *Pa-ga-ti-thar-ga-ra* crystal blue ocean in the west and *Pi-ta-thar-ga-ra* the golden ocean on the northern side of Mount **Meru**. Large islands occupied the middle of each ocean.

Pubbavideha, the half-moon shape island of the east ocean is 7,000 yojana in length.

Jambudîvpa, the cart shape island of the south ocean is 10,000 Yojana across.

XLI The drawing in the manuscript does not show the tripod of the three mountains supporting Mount Meru mentioned in the translation (see Figure 16).

XLII It is noted that many of the figures given for dimensions in the manuscript are slightly at odds with those of King Lithia's work. As an example; Mount Meru is now assigned a height of 86,000 yojana above the surface of the ocean. The Trai Phum has Mount Meru reaching to a height of 84,000 yojana above the sea and the *Abhidharmakośabhāṣyam* has Mount Meru soaring to 80,000 yojana into the sky.

XLIII The reference to scholars is as translated, and is not an augmentation.

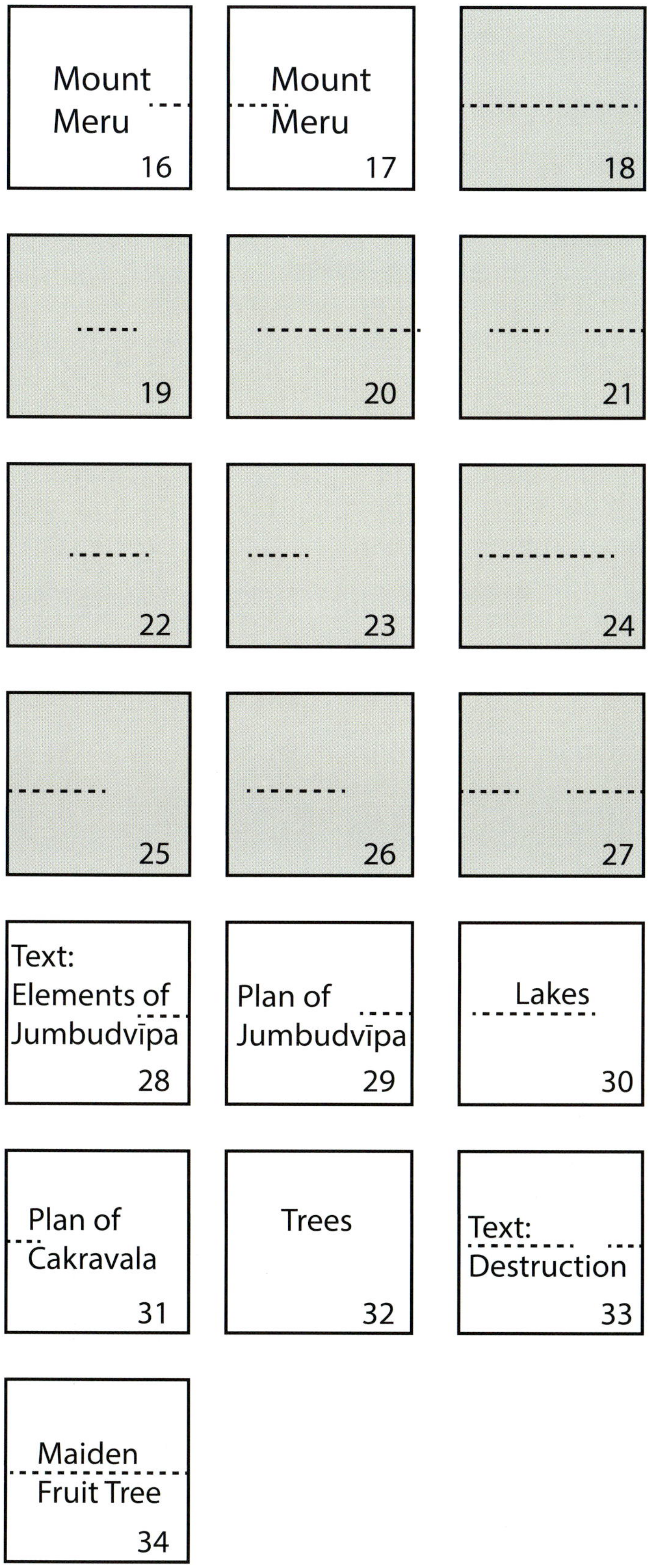

Mount Meru 16	Mount Meru 17	18
19	20	21
22	23	24
25	26	27
Text: Elements of Jumbudvīpa 28	Plan of Jumbudvīpa 29	Lakes 30
Plan of Cakravala 31	Trees 32	Text: Destruction 33
Maiden Fruit Tree 34		

Aparagoyāna, the full moon shape island of the west ocean has a diameter of 8,000 Yojana.

Uttarakuru, the stone slab shape island of the north ocean has a length of 8,000 Yojana.

Each main island has 700 small islands of the same shape as their main island. Altogether, there are 2,000 small islands in the four oceans which are all surrounded by the mountain range of the universe, which is 8,200 yojana high above the ocean.[XLIV] As an example, set a round tray afloat a pond and you will find, the floating tray was like the universe, and the pond was like the water body which supported the universe. The number of universes were uncountable, like several trays floating on the pond with each rim touching each other, as infinite universes stood endlessly in the space. Among the triangular space unoccupied by the universes is the **Lokanta** Hell of total darkness. The sufferers crawled up and hung like bats on the wall of this universe; they struggled but fell back into the darkness to suffer like dough being fried in deep hot oil. (T)[XLV]

✲ ✲ ✲

The artists and writers of this manuscript devoted seven manuscript pages to the exploration of the Cakravāla and the continent of primary interest: 'Jumbudvīpa'.

The sketch shows manuscript relationships between Mount Meru and the manuscript pages which deal with Jumbudvīpa and the Cakravāla; the intervening are the pages on *Preta, Asura* and the hells.

Interestingly, the manuscript avoids the possibility of setting the stage for the discussion of the Jumbudvīpa by introducing the plan for the Cakravāla after the plan for Jumbudvīpa. Normally, it would be expected that the universal type plan would proceed the parochial part, i.e. the Cakravāla first then Jumbudvīpa. The path of the manuscript used for the presentation of cosmography will also be followed after the two page introduction of Mount Meru.

XLIV Meaning not clear. Believe it is a reference to the Cakravada iron mountain range which contains the great ocean of the Cakravāla.
XLV (T) = translation

Figure 29. Mount Meru and its surroundings (Manuscript pages 16 & 17)

Mount Meru: The Great Cosmic Axle of the Cakravāla

The summit: The summit of Mount Meru is the site for Tāvatimsa Heaven. The great mass of the mountain column provides the homes of the four guardian gods of the cardinal corners of the Cakravāla

The Celestial Orbs: Suriya-Devaputta, the sun god seen on the left of Mount Meru and Candra-Devaputta, the moon god, is on the right.

Sattaparibhanda: The term for the seven mountain ranges surrounding Mount Meru and the Sīdantara Sea which becomes more shallow as it moves away from the great massif.

The Great Ocean

Mount Meru

Mount Meru is a highly decorated cosmic column in bright colors with a "fish scale" type motif. It is depicted over the expanse of four manuscript leaves. The heaven of the four kings of the cardinal quarters (*Cātummahārājika*) is housed within the column.

Figure 30. The Orbits of the Sun and Moon (Manuscript page 17)

The Celestial Orbs

> The sun and moon are shown in orbits at the height of the highest of the mountains surrounding Mount Meru

Suriya-Devaputta – the sun god is shown with its identification of a rabbit with four numbered satellites. These satellites indicate the existence of additional planetary orbs.[152]

Candra-Devaputta – the moon god with its peacock identification.

Both the sun and moon orbit at the same level of the Yugandhara Mountain Range. This range is the highest of the ranges surrounding Mount Meru. The small satellites of the sun and moon may be representations of the five planets and Rahu, the eclipse. Together these form the eight astrological compartments.[153]

Sattaparibhanda: The Seven Mountain Ranges Surrounding Mount Meru[XLVI]

Seven mountain ranges surround Mount Meru. These are shown as large candlestick-like columns, each surmounted with a small temple-like structure. In keeping with the extremely structured nature of the cosmos, the seven surrounding mountains diminish in orderly size according to their distance from Mount Meru. The sea bed rises uniformly as the distance increases from Mount Meru. The surrounding mountain ranges beginning with the highest and closest to Mount Meru are:

Yugandhara Mountain Range
Īsadhara Mountain Range
Karavika Mountain Range
Sudassana Mountain Range
Meemindhara Mountain Range
Vinataka Mountain Range.
Assakaṇṇa Mountain Range

XLVI See Sattaparibhanda Golden Mountains and The Sīdantara Sea in Part 1, for a description of the mountains and the seas.

At the Base of Mount Meru

The black chambers at the base of Mount Meru lie just above the Asura palace and beneath the chamber of Lord Virūpākkha, who rules all the devatā on the western side syambolizes by the garuda and nāga,[154] who fall into the domain of Lord Virūpākkha. This chamber is in turn above three black vault-like rooms which are over the Asura Palace and whose spire pentrates through them and which may indicate the desire of the Asura's return to the heavens above. However, the reason for the three black chambers has yet to be discovered.

Figure 31. The Black Chambers

The Great Ocean and the Ānandā Fish

The great ocean in which Mount Meru rests is teaming with fish and other creatures. The king of the fishes, a giant called Ānandā, is responsible for keeping the ocean in working order. This vignette, extracted from the manuscript, shows other inhabitants of the great ocean including that of a confused European type fish, and fish with elephant and cow heads. Clearly the artists enjoyed this part of the work.

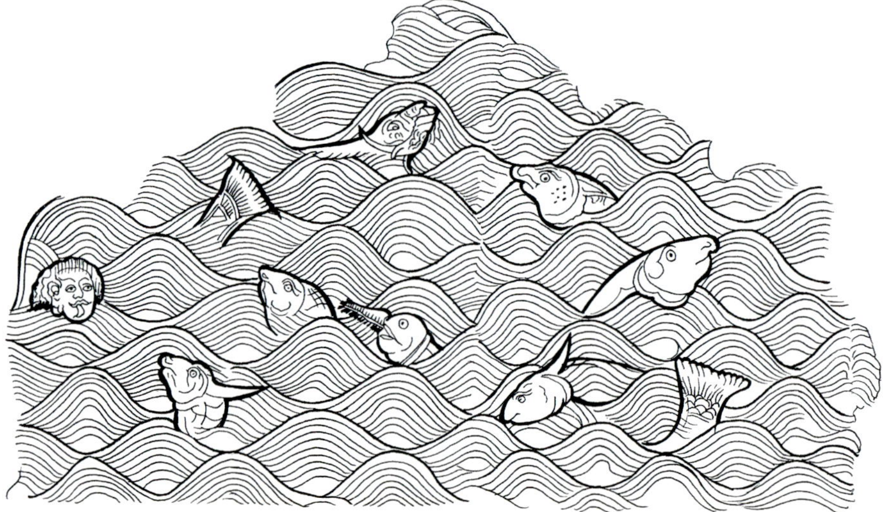

Figure 32. The Great Ocean and the Ānandā Fish – A Vignette

Jumbudvīpa's Cosmography: Translation[XLVII] of Manuscript Page 28

✲ ✲ ✲

In the south continent Jumbudvīpa the Himavanta mountain range has five mountains which surrounded Lake **Anotatta**. All these mountains have the same height and width of 50 yojanas. And under a shining full moon, these mountains shine like burning flames.

1. **Thudatthana** Mountain is made up of pure gold.

2. Seiktara Mountain. is made up of various kinds of precious stones.

3. Kala Mountain is made up of cat's eye moonstones.

4. Gandamardana Mountain is full of fragrant wood, spices and other medicinal plants, such as caramel, sandalwood, clove, nutmeg, cobra's saffron, ebony, beetle leaf tea leaf and other medicinal plants. The fragrant scent of the flowers, leaves, barks from this mountain gave it the name Gandhamadanakuta.

5. **Kaelarthaba** Mountain is made up of pure silver.

Anotatta Lake receives light from the sun and moon, only when they travel from north to south. **Anotatta** Lake receives no light when the sun and moon passes in other directions.

Anotatta Lake is a place of paradise.

There is a descending staircase made up of yellow arsenic and other minerals and studded with precious gems. There are teak trees bearing flowers of gold, crystal and ruby colors. The water was clear and free of any impurity. Only Gods, Bodhisattvas and powerful Nats and giants could drink and bathe in this water.

Lake **Anotatta** has four outlets:

> In the East, water flows from the mouth of the lion in clock-wise direction three times around the lake before it flows into the eastern ocean. Along its course reside devils, ghosts and giants. The water does not mix with other outflows from the lake.
>
> In the North, water flows out from the mouth of a horse to the northern ocean.
>
> In the West, water flows out from the mouth of an elephant to the western ocean.
>
> In the South, water flows out of the mouth of a bull, and goes around the lake three times in a clock-wise direction and then turns south, where it meets a 60 yojana long stone plate. When the water first flowed the initial impact of the current on the stone plate resulted in a strong three Gar-wave three-quarters of a Yojana in diameter, flying 60 yojana high before it drops down on a stone plate named Tiyaggala. The impact broke the stone plate and a lake with the same name was formed with a depth of 60 Yojana. From Tiyaggala Lotus Lake water flows through a 60 yojana long underground tunnel, where it was attacked by Bi-la-si-thor-wis-sa[XLVIII] and the breaks into five different rivers which all of which flows into the southern ocean. The five fingers-like rivers are the Ganga, Yamuna, Irrawaddy and Tha-ra-bu.[XLIX]

XLVII Note: The translation is as it was written, creating confusion between the two lakes. The first part up to the Eugenia trees, could be for the Saddan Lake alone and the latter part for Mandakini Lake.

XLVIII A mythical animal.

XLIX Only four rivers are named!

Among the five mountains that surround Anotatta Lake, the Gandhamadanakuta Mountain is the place most suitable for Pacceka Buddhas. There are three caves, one of gold, one of silver and one of ruby. Near each cave stand a 100 yojana high Mañjūsaka Tree, divine flower tree, which is bejeweled by various kind of plants and flowers with fragrant smell, taste and beauty.

Besides Anotatta Lake, there are 6 more lakes.

On the west side of lake Saddan, is a mountain of gold, which, in the rainy season creates golden waves 12 yojana high each in the lake. This is the time when Saddan, the king of elephants and his 80,000 followers came to enjoy the waves. In the summer months, they relaxed among the 8,000 prop roots of the great Banyan tree and enjoyed the cool breeze flowing from the lake. In the Mandakini [?] lake, there is a body of dirt free, crystal clear water twelve yojana wide. It is surrounded by a belt of white lotus that is one and one-half yojana in width. Milk oozes out from each segment of the roots which are the size of a pyee[L] The milk flows on top of the leaves, and when cooked by the sun light, it becomes a reddish honey like substance, which is tasty and nutritious. After this white lotus belt comes a colorful lotus belt of white, red and brown lotus. It is again surrounded by wild rice fields, gardens of cucumber, sweet gourd, pumpkin and then various vines, then sugarcane fields, each stalk the size of a betel nut palm. There is a large stone slab, where the elephants stamp the cane to extract juice for their own consumption. The remaining juice on the stone is cooked by sunlight and becomes a delicacy, which the elephants made as offerings to the Buddhas, the enlightened ones, and the arhats. From there comes the forest of bananas, each fruit the size of an elephant's tusk, Jackfruits, each the size of a clay barrel; mangoes, each the size of a Pyee basket; then comes the forest of Eugenia trees.

Around **Mandakini Lake**, there are no fruit bearing trees that cannot be eaten or are harmful. The lake itself is fifty yojana in length and ten yojana in width. In the **Himavanta** forest, there is a mountain called Thu-wun-na-pap-pa-ta[LI] and **Mandanakini** situated on the north side of this mountain. This lake also has trees bearing fresh, clean fruits and fragrant flowers on all four sides of the lake. There are Zaung-dan[LII] and step ways. It is a place where **Indra** and other deities come to enjoy themselves. On the south-east side of the lake, there is Nan-dar cave, 50 yojana deep and 10 yojana wide, filled with gems and precious stones. At the mouth of this cave stands a banyan tree, with the girth of five yojana and, a height of 14 yojana. The tree has five main branches each six yojana long and 8,000 roots which serve as propping for the massive tree. Saddan and other elephants sometimes come to rest among these great propping roots. It is also the site where Ashin Mahakassapa[LIII] spent his last 12 years, meditating and practicing. Taking turns, the elephants served him with food, drink and other necessities until he departed to Nibbana. The other four lakes are also surrounded by various trees and plants bearing fruits and flowers. (T)

L Believed to be of three liter capacity.

LI The Pali, or other name not identified.

LII A covered type staircase to pagodas and temples.

LIII One of Buddha's ten great disciples and later, upon the death of Buddha, took over the leadership of the Sangha. He is also known as Mahakāśyapa, i.e. the great Kāśyapa.

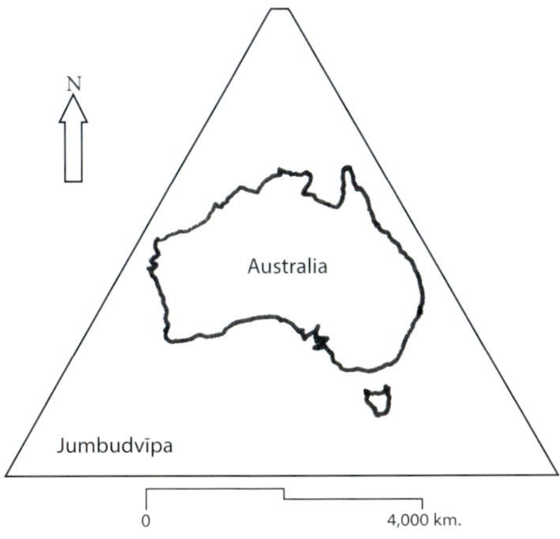

Figure 33. Jumbudvīpa and Australia with their plans to scale

Jumbudvīpa: Plan and Major Elements

Jumbudvīpa continent has soaring mountain ranges, sea size lakes, great forests, and large, strong running rivers. The geological features of Jumbudvīpa are described in detail by King Ruang[155] but much less so in the *Abhidharmakośabhāṣyam of Vasubandhu.* This Burmese manuscript, which illustrates and describes the total cosmos, depicts only some of Jumbudvīpa's main characteristics. Jumbudvīpa, the southern continent could be styled as 'our' world.[156] The approximate area of the Jumbudvīpa Continent is 28,000,000 square kilometers or 10,800,000 square miles. The earth's largest ocean continent, Australia, is shown in Figure 33 to scale. The configuration, an almost equal lateral triangle, is according to *Abhidharmakośabhāṣyam of Vasubandhu.*[157] See Table 2 for comparisons of the land masses of the other continents.

Figure 34. Jumbudvīpa's Plan (Manuscript page 29)

The plan and location of the important central elements of Jumbudvīpa are depicted in the manuscript with minimal detail and great innovation. The plan's geographic places and three main elements are shown in the following figures.

Figure 35. Specific Named Jumbudvīpa Locations

No.	Jumbudvīpa Locations	No.	Jumbudvīpa Locations	No.	Jumbudvīpa Locations
1.	Anotatta Lake	9.	Ruby Cave	18.	Mahi River
2.	*	10.	Mañjūsaka Tree	19.	Sarabhū River
3.	*	11.	Bawoodagakinga**	20.	Oomakinga**
4.	Ganga River	12.	Kaelarthaba Mountain	21.	Haylaga lake**
5.	*	13	Bagatakinga***	22.	Tiyaggala Lotus Lake
6.	Gandhamadanakuta Mountains	14.	Karakinga**	23.	Thudatthana Mountain_
7.	Gold Cave	15	Animal Shaped Mountain	24.	Sakdara Mountain **
8.	Silver Cave	16.	Yamunā River	25.	Kalakuta Mountain
		17.	Aciravatī River		

Major Elements

The three major components of Jumbudvīpa are:

1. **The Buddha with Devotees**
2. **The Rivers and Lakes**
3. **The Mountains Surrounding Lake Anotatta**

Many more items described by King Ruang and others are not shown in this Burmese manuscript. Additionally, the manuscript has a number of otherwise unidentified elements. Various translations for the locations and specific elements are presented in Appendices A and B.

Figure 36. Major Elements of Jumbudvīpa's Plan

The Buddha with Devotees

The Buddha is seen in an awkward pose with the right arm in possibly a variant of subduing Māra pose and the left arm is in a yet unidentified gesture, a position whose meaning is unclear.[LIV] Kneeling before the Buddha, are crowned adoring devotees in patterned gowns, one devotee is holding flower offerings the other a whisk. Between the Buddha and his devotees is a Burmese glass-and-lacquer offering vessel holding holy water. This portrayal is to be found in most manuscripts which deals with Theravāda cosmology. The Buddha is seated on a golden narrow waisted lotus throne. The southern border has a protecting wall which morphs into sheltering mountains as it goes from right to left.

AN OFFERING VESSEL

Burmese Offering Vessel (author's collection): This offering vessel is used to present food and gifts to monks; it was presented to a monk by a lay person to accumulate merit. This kind of vessel is also used by senior monks to bestow blessings on laity by the use of a whisk dipped in holy water inside the glass bowl. The vessel is composed of lacquer on coiled woven basketry; relief is built-up by successive layers of lacquer. The decoration of this example is geometric, with small pieces of colored glass inserted into the lacquer.

LIV A number of gestures are described and illustrated in K. I. Matics' book *Gestures of the Buddha* and Professor Khaisri Sri-Aroon's *Buddha images in Thailand (Siam)* but the one in the figure is not among them. The closest are two shown in Suwat Saenkattiyarat's book *Iconography of the Buddha* which has a conquering Māra illustration but with the left hand resting palm up on his lap the other illustration is the ordination of The Buddha's first disciple with the left hand palm resting below the knee. The manuscript's illustration shows the Buddha's right hand touching the earth, or subduing Māra position.

Rivers and Lakes

In the cosmology of the Cakravāla there is a great mythical body of water at the center of the Jumbudvīpa called Anotatta Lake. This lake is the source of four great rivers. Only one river, the Ganga (Ganges) is named; the names of three other rivers are not given either on the manuscript drawing, or in the accompanying text.[LV]

Figure 37.
The Anotatta Lake and its rivers

The lake in this manuscript is shown in a circular form. It is also sometimes depicted as a square with 50 yojana on each side (2,500 square yojana, or 40,000 square kilometers in area). Sadakata suggested that this lake may be a transformation of the diminished Lake Mānasarowar in Tibet.[158] The nomenclature for the lake is a compromise conforming both to the translation from Burmese (Appendix A) and as presented in the Reynolds and Reynolds translation.[LVI]

The following sketch extracted from the manuscript is to explain the source of the rivers. The rivers emerge from the lake via rocky shaped outlets that resemble animals' heads.[159]

> The river from the mouth of a lion flows east into the ocean.
>
> The river from the mouth of a cow (also described as a bull or ox) flows west into the ocean.
>
> The Ganges River (Gaṅgā River) flows south from the mouth of an elephant, and divides into five branches which goes through the plains, towns and farms of Jumbudvīpa into the ocean.
>
> The river from the mouth of a horse[LVII] flows north into the ocean.

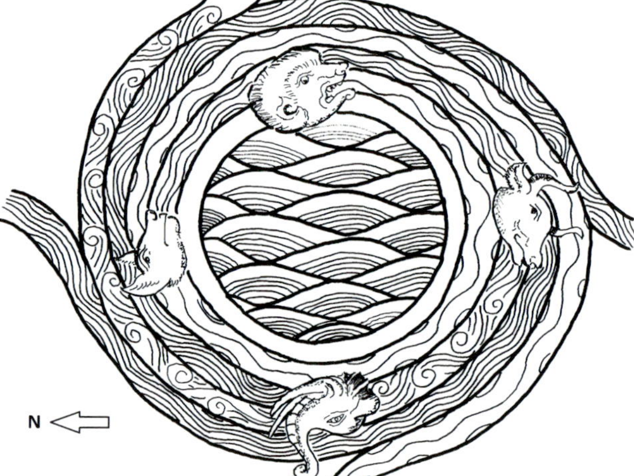

LV Names for the rivers were also not found in the Reynolds and Reynolds and the Thai National Team's translations of the '*Traibhumikatha*'. Akira Sadakata does name them, Indus, Oxus and Śitā, and makes good argument for his names (see Sadakata 2009, p. 31).

LVI The spelling 'Anavatapta' Lake may be found in other works.

LVII Some of these mythical rivers; especially with the use of Akira Sadakata's nomenclature may have earthly counterparts. The Indus River is the main river draining Pakistan. The Ganges River flows from Pakistan through Northern India. The Oxus River, identified by Sadakata (2009, p. 32) as the Amu Darya River, which forms part of the northern boundary of Turkmenistan (see p. 222 Geographica 2004) flows through Turkmenistan/Uzbekistan, It is sometimes called the 'Tarim' and is thought to be the great Yellow (*Huang He*) River of China. Sadakata suggests that the Śitā River may be the Yarkland (Sadakata 2009, p. 32.) The Yarkland River flows southwest from the Himalayas near Aral City in Western Xinjiang and joins the Aksu River to form the Tarim River (Wikipedia. http://en.wikipedia.org/wiki/Tarim-River (accessed12/27/2013).

The Himavanta Lakes

The following page of the manuscript shows six of the seven Himavanta lakes. Lake Anotatta, the seventh and the most important, was already visualized in the plan for Jumbudvīpa. The six remaining lakes are all 50 yojana deep and 50 yojana wide. The Himavanta Lakes shown in the following figure are (l to r):

1. Kannamunda Lake
2. Kunala Lake
3. Rahtakāra Lake
4. Sihappāta Lake
5. Saddan Lake
6. Mandakini Lake

The drawing of the sprig of green to the right of the lakes may represent Anotatta Lake. These fragrant flowers, bark, and trunks are found in the Gandhamadāna mountain range-one of the five mountain ranges that surround Anotatta Lake. (See Figure 35 for specific named Jumbudvīpa locations.)

Each of the six lakes is 50 yojana wide. Only one lake, Saddan, enjoys an extensive description in the manuscript. (See the translation of Manuscript page 28 above.)

Figure 38. The Lakes of the Himavanta (Manuscript page 30)

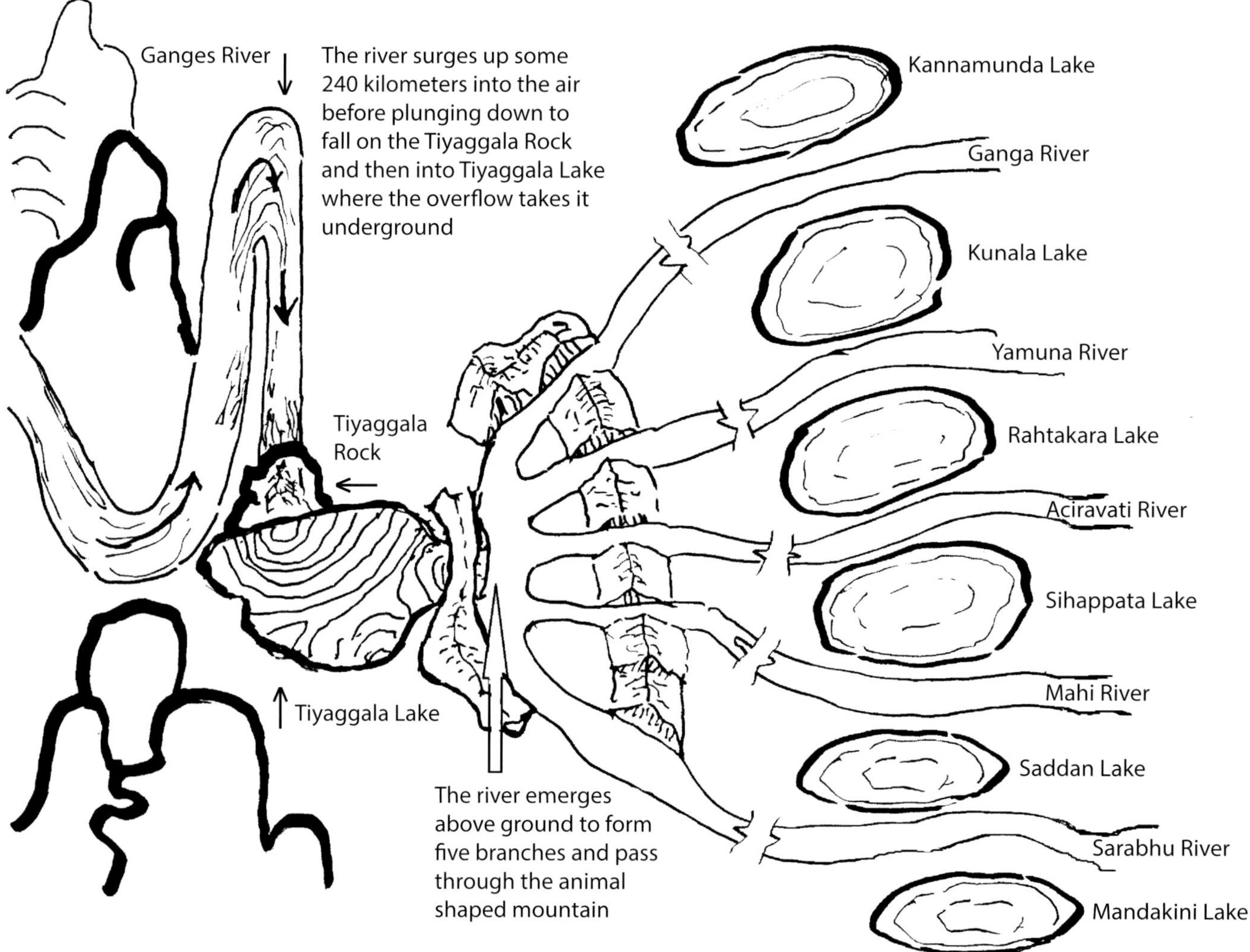

Figure 39. An Extrapolation of the Ganges River System (expanded from Figure 38)

Figure 39 goes beyond the manuscript's illustration to show the locations of the six additional lakes, to clarify the river branches, and to explain the great loop in the Ganges River shown in the plan. Anotatta Lake nonetheless is the sole source of the rivers. The order of the naming of the lakes is arbitrary and is in serial order, north to south.

The Ganga River undergoes a number of name changes based on its activities as it flows towards the ocean. At first it is called the *Avaṭṭagangā* then the *Kanhagaṅgā.* When shooting up into the sky it is called the *Ākāsangaṅgā.* Then it is called the *Bahalagaṅgā* when going through the rocks. Finally it is called the *Ummagga-gaṅgā* when going underground.[160] These rivers are also described as having great volumes of water which differ at times during their passage to the ocean.

The Mountains Surrounding Anotatta Lake

The Anotatta Lake is circled by five mountains each 200 yojana (800 kilometers) high.[161]

1. Thudatthana Mountain
2. Seiktara Mountain
3. Kala Mountain
4. Gandamardana Mountain
5. Kaelarthaba

The manuscript (1842 C.E.) which preceded the English language translations of King Lithai book by more than a century, leaves little question that King Lithai's *Traibhumikatha* was one of the basis for this Burmese manuscript. See the following tabulation:

Manuscript (Transliteration)	Three Worlds According to King Ruang[162]	The Story of the Three planes of Existence[163]
1. Thudatthana Mountain	1. Sudassana Range	1. Sudassanakuta Mountain
2. Seiktara Mountain	2. Citta Range gems	2. Chitrakuta Mountain
3. Kala Mountain	3. Kāla Range	3. Kalakuta Mountain
4. Gandamardana Mountain	4. Gandhamadana Range	4. Gandhamadanakuta Mountain
5. Kaelarthaba	5. Kelāsa Range	5. Krailasa Mountain

Gandamardana Mountain (3A in Figure 37): Fragrant Mountain[164] is said to glow like a burning ember during the full moon and is rife with caves; it supports fragrant vegetation. The mountain has silver, ruby and gold caves. These caves are the homes of the Pacceka Buddhas,[165] arhats who achieved self-enlightenment. They became enlightened without ever hearing, or even knowing about, the Dharma, or a doctrine. The Pacceka Buddha reaches the understanding of the Four Noble Truths by himself and without a teacher. The Pacceka Buddha cherishes solitude and is not talkative, hence he does not become a teacher. After many eons the Pacceka Buddha "utters an aspiration before a Perfect Buddha" and achieves his goal.[166] The caves shown on Jumbudvīpa are the habitats of the Pacceka Buddhas. The material used for the caves, gold, silver and ruby, is indicative of the respect and esteem in which these holy men are held.

Thudatthana (Sudassanakuta) Mountain (Item 23 in Figure 35) From the Traibhumikatha, by King Lithai, we learn that the mountain is made of gold and borders the lake like a wall and curves towards the lake like a raven beak.

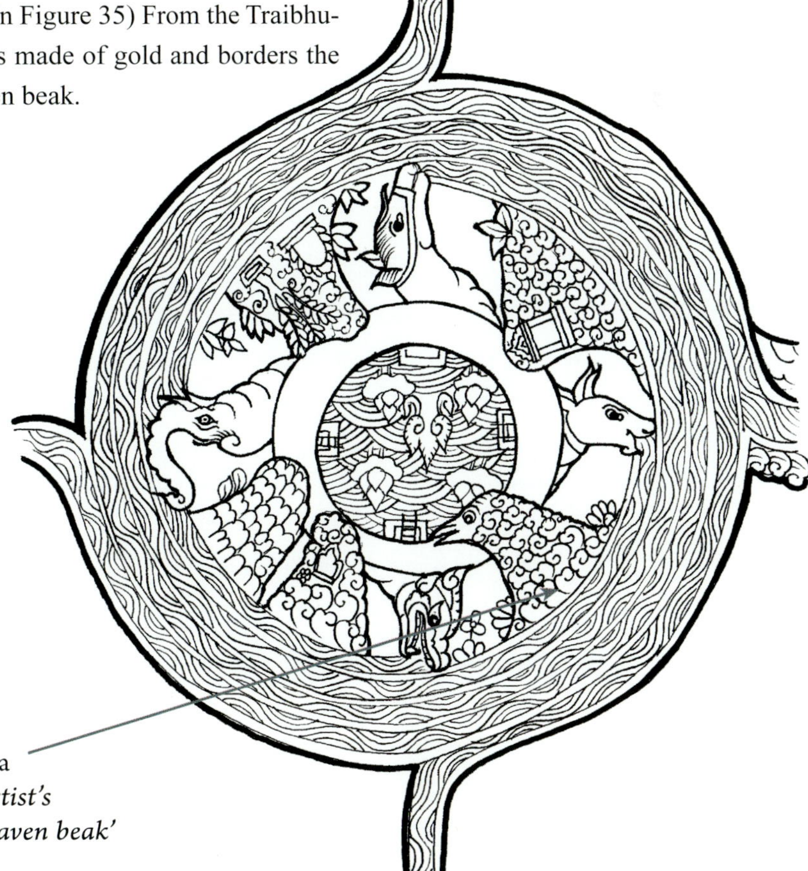

The sketch, extracted from Trai Phum Book: Ayutthaya Manuscripts–Thonburi Manuscripts *shows another artist's interpretation of Thudatthana (Sudassanakuta the 'raven beak' mountain).*[167]

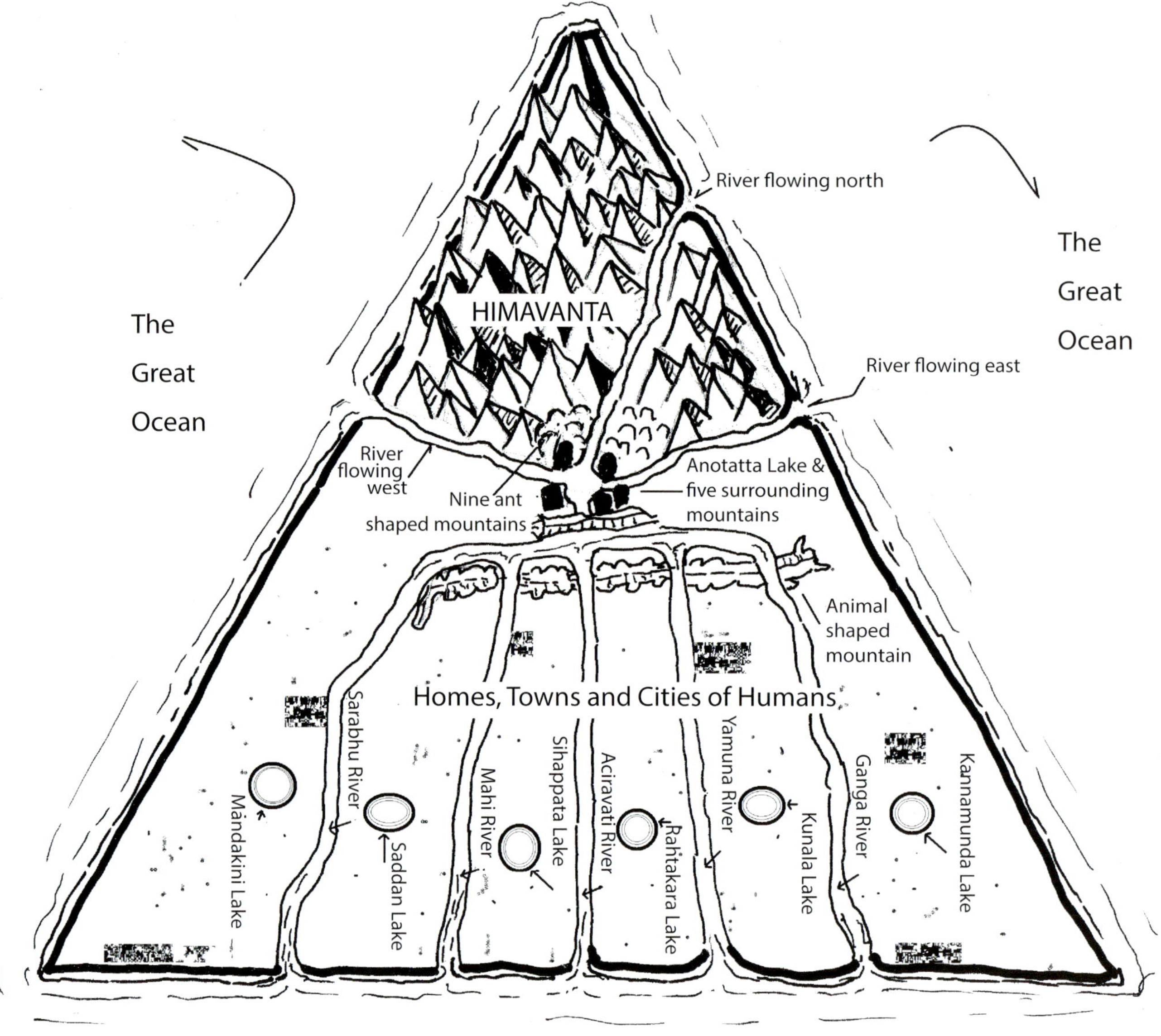

Figure 40. A Visualization of a possible geography of Jumbudvīpa

Possible Geographic Aspects of Jumbudvīpa

The geography of a mystical land, resting in a world of transcendence requires specifics, or imagination, or both. In King's Lithai's writing, some specifics are presented, but descriptions of the overall sizing and physical arrangements of Jumbudvīpa leave much to be desired. For the configuration and dimensions of Jumbudvīpa, we turned to the fifth century C.E. writings of Vasubandhu,. For other details of Jumbudvīpa we find those in the writings of King Lithai in this manuscript.

Cakravāla

The Plan Of The Cakravāla

Cakravāla main geographic components are the great mountain Meru, the seven mountain ranges surrounding Mount Meru, and the four island continents and their satellites at the cardinal points of the universe. These are shown in following figure. See Figures 13 & 14 for other details.

Figure 41. Manuscript Plan of the Cakravāla (Manuscript page 31)

Mount Meru is in the center depicted as a large black dot surrounded with seven annular mountain ranges, shown as black, target-type, rings all of which rest in the great ocean filled with giant fish. The red circle represents the Cakravāda iron mountain range which contains the oceans waters. The blue circle represents the continuum of space surrounding the Cakravāla.

As noted; the continents with their islands all have prescribed shapes:

North of Meru is Uttarakuru which is square

South of Meru is Jumbudvīpa which is shaped like a cart

East of Meru is Videha which is shaped like a half moon

West of Meru is Godaniya which is round

Trees and Other Features

The manuscript contains a drawing of sacred trees which are identified with the various continents, islands and one heaven. The drawing also shows six Cakravāla world systems, each with four islands; between them is the extremely cold Lokanta Hells (see Part 3, section on 'hells,' for a description).

There is also a tree – the 'Maiden Fruit Tree' – with youths and maidens hanging from the branches.

Trees of the Cakravāla[168]

1. South of Meru is Jumbudvīpa and the Rose Apple or Eugenia tree, *Zanbu thapye,* from which the continent is named.

2. West of Meru is Aparagoyāna with the Hteinpin tree.

3. North of Meru is Uttarakuru with the Padether tree.

4. East of Meru is Pubbavideha with the Kukko tree (rain tree).

5. North East of Sudassana City, some 20 Yojana distance, in Tāvatimsa Heaven, is the Pinle Kathit[LVIII] tree.

6. Under Mount Maru, in the state of Asura, is the Thakhut (trumpet flower) tree.[LIX]

7. The state of Galuna, is home to the Latpan tree (red silk cotton tree silk cotton or kapok) tree, sometimes mistakenly called "the flame of the forest." *Bombax ciebam*[LX] (T)

Figure 42. Trees of the Cakravāla and the Lokanta hells (see Figure 27) (Manuscript page 32)

The Mañjūsaka tree while not illustrated on this manuscript page is shown in the extreme northwest of Figure 34, Jumbudvīpa's Plan[169] (Manuscript page 29) and is said to be 100 Yojana (400 kilometers) high.

The Cakravāla's Destruction

All Cakravālas are subject to cycles; cycles of development and cycles of destruction. The cycles of destruction occur periodically and are either totally or partially destructive. Destructive forces are fire, wind and water. After immeasurable periods of time recreation occurs and life begins again.

LVIII Family – *Leguminosae / Papillionideae.*

LIX Both are of Family – *Bigononiaceae.*

LX Family – *Bombacaceae.* The Female Thakut – Tui or mangrove trumpet tree, *Dolichandrone spathacea.* The Male Thakut – trumpet flower, yellow snake tree, *Stereospermum personatum.*

Translation of Manuscript Page 33

✲ ✲ ✲

Thus the universe was formed and was called the hundred-thousand Ku-dae universe. (each Ku-dae is equal to 10 million). At each beginning of the universe, the sacred site where the various Buddhas will be enlightened, the bejeweled throne and the Bodhi tree are the first to be realized and the last to be destroyed.

Among the infinite universes, our universe is the only place where Pacceka Buddhas, [enlightened monks] the Sa-kya-wa-tae-min[LXI] [the emperor of the universe] are to be found and where Ya-han-thar exists. They do not exist in other universes.

The universe was formed and destroyed through a period of time called Athincheya-Kappa which is composed of the following four Kappas:

Tharn-wa-da-Kappa

Tharn-wa-da-Htar-yi-Kappa

Wi-wa-da-Kappa

Wi-wa-da-Htar-yi-Kappa

When all the above four Athincheya-Kappa are combined they are called a Maha Kappa.

Among the four Kappas

In a fire destroyed Tharn-wa-da Kappa, the period beginning with the first rain of fire until its extinguished, was called a Tharn-wa-da-Htar-yi Athincheya Kappa. After which a constructive rain will fall.

From there onwards to when the sun, moon and constellation occur, was called Wi-wa-da Athincheya Kappa.

From there onwards until the disappearance of Sun, moon and constellations, this period was called the Wi-wa-da-Htar-yi A-thin-che-ya Kappa.

All these four A-thin-che-ya Kappa combine and were called a Maha Kappa. The duration of each A-thin-che-ya Kappa is the same.

This present world will be destroyed by fire.

For a water-destructed world the period from the first destructive rain to when the water vanishes.

For a wind-destructed world, the period between the first wind storm to its termination l begins with the Tharn-wa-da -A- thin-che-ya Kappa.

This world will be destroyed only by the three elements, fire, water and wind for several times.

The world will be destroyed –

Seven times by fire,

Eight times by water

Sixteen times by rain and wind

Destruction by fire will reach and destroy the first heavenly Brahmā level

Destruction by water will reach the 2nd level and destruction by wind it will reach the third level. The levels above them will not be affected. (T)

✲ ✲ ✲

LXI Mahacakkavattiraja (Pali) – a universal emperor. Sa-kya-wa-tae-min –'min'is Burmese for a kingly rank.

TABLE 6. DESTRUCTION BY LEVELS with the Various Forces of Destruction

Level	Pali Name[170] and Condition	Forces of Destruction		
		Fire (7 times)	Water (8 times)	Rain & Wind (16 times)
1	**Nevasaññā-nāsaññāyatana** – Neither-perception-nor-non-perception			
2	**Ākiñcaññāyatana** – Nothingness			
3	**Viññānaññcāyatana** – Infinite consciousness			
4	**Ākāsānaññcāyatana** – Infinite Space			
5	**Akaniṭṭha** – Clear-sighted devas			
6	***Sudassī*** – Clear sighted Devas			
7	***Sudassā*** – Beautiful Devas		Heavenly levels spared destruction	
8	**Atappā** – Serene Devas without troubles			
9	***Avihā*** – Devas not falling from prosperity			
10	***Asaññasatta*** – Devas without perception			
11	***Vehapphala*** – Devas who are fruitful			
12	**Subhakiṇṇā** – Devas of unlimited splendor			
13	***Appamāṇasubhā*** – Devas of unlimited beauty			
14	***Parittasubhā*** – Devas of limited beauty			
15	**Ābhassarā** – Radiant Devas			
16	**Appamāṇabhā** – Devas of unlimited splendor			
17	**Parittābhā** – Devas of limited splendor			
18	***Mahābrahmā*** – Mahabrahmā -The Great Brahmā			
19	***Brahmaparohita*** – Deva priests & ministers of Mahabrahmā			
20	**Brahmapārisajja** – Devas in the retinue of the Mahabrahmā		Levels destroyed	
21	***Paranimmita-vasavattī*** – Devas who find pleasure by others			
22	***Nimmānarati*** – Devas who enjoy their own works			
23	***Tusita*** – The birthplace of *Bodhisattvas* (The future Buddhas)			
24	***Yāmā*** – The level of Yāmā Devas			
25	***Tāvatimsa*** – The home of Indra (Sakka) and 32 Devas			
26	***Cātummahārājika*** – The four kings of the cardinal quarters			
27	***Manussa*** – The level of humans			
28	***Asura*** – The level of fallen devas and demons			
29	***Preta*** – The level of the hungry ghosts			
30	***Tiracchāna*** – The level for all animals			
31	***Niraya*** – The hell world of the damned			

Figure 43. Maiden Fruit Tree (Manuscript page 34)

The Maiden Fruit Tree

The maiden fruit (thu-yong-thee) tree is rare and grows wild in the deep, remote areas of Himavanta. It bears fruit resembling the shape of human beings. The male fruit is a boy of 20 years and the female fruit has the beauty and shape of a 15-year-old girl. All are well clothed and when they ripen they fall to the ground and are eaten by birds leaving only an inner core which resembles a human skeleton. (T)

Figure 43 is the last drawing of the series of the cosmos and precedes those of the arhats. A question arises as to why this drawing is the last one in the pages of the cosmos, and even why this drawing is in the manuscript. This illustration shows both maidens and young men dangling from their heads in a tree in full blossom. At the base of the tree is a rishis hermit. The maiden-fruit tree is found on the Jumbudvīpa continent[171] in the Himavanta.[172] Normally, the description would indicate that the tree bears fruit shaped like maidens who have just become 16, when men fall in love with them. When the fruit (maidens) become ripe, they fall to the ground and birds eat them[173] leaving exposed an inner core which resembles the human skeleton.[174] This standard description for the Maiden Fruit Tree found in other works is at varance with both the illustration and description in this manuscript.

The following is sourced from a learned scholar and teacher.[LXII]

The drawing also shows three figures which are not attached to the tree but are seen flying or hovering in the tree. The upper figure on the left holding a maiden fruit is *Sawvgar;* the one on the right is *Phattaavigar.* They are brothers who reside in palaces in the sky between the earth and heavens. The third one below *Sawvgar* is *Naaraphatti* and has a place on a mountain top. He is capable of changing his appearance at will. All three have small crowns and cannot gain Nibbana because of past deeds. Their main focus is stealing and co-habiting with the maiden fruit for the period of time before the fruit spoils.

RISHIS

In Southeast Asian paintings are to be found male figures sometimes clothed in animal skins with strange looking headgear. These are rishis ascetic hermits. Rishis are holy men; ascetics, seers, shamans, saints, poets or hermits. Rishis speak the truth while others falter, because they exist on a higher plane of consciousness. According to legends, rishis are found in the remote reaches of forests and mountains and are known for their medical cures based on herbs and other natural items. Rishis have ancient roots and rose to prominence in early Hinduism some 4000 years ago. The Vedic hymns were revealed to seven rishis.

In the *Trai Phum Book: Ayutthaya Manuscripts–Thonburi Manuscripts* there is only one drawing of a 'maiden fruit tree' in the two volumes of antique manuscripts. The following black and white drawing extracted from the book shows *Sawvgar, Phattaavigar* and *Naaraphatti* harvesting the fruit and flying away (seemingly confirming the foregoing?). The drawing also shows the fruit emerging feet first and hanging by the head for seven days until harvest or it falls to the ground.[175] The text accompanying the drawing identifies the tree as a '*Nareephon* Tree*'* or 'fruit women'.

Figure 44. A Maiden Fruit Tree from a Thai Manuscript[176]

LXII An 'Adjhan Wat' (Temple Professor) of a temple in Chiang Rai Province, Thailand, formerly Abbott of a Burmese Temple with 25 years' service in the Sangha.

PART 5

The Arhats

Introduction

> The expression arhat literally means one who is worthy of veneration. In early Buddhism the word applied to a monk who had attained enlightenment, but later it came to signify a specific group of persons who were especially selected by the Buddha to spread his message all over the known world.[177]
>
> The arahat is the ordinary hero of the Buddha's Way as it was first enunciated.[178]

Theravāda Buddhism regards the doctrine that guides the arhats as the original teaching of Buddha. The doctrine deems that the Monk-Arhat must isolate himself from the distractions of worldly things and concentrate on meditation.

> For a disciple thus freed, in whose heart dwells peace, there is nothing to be added to what has been done, and naught more remains for him to do. Just as a rock of one solid mass remains unshaken by the wind, even so, neither forms, nor sounds, nor odors, nor tastes, nor contacts of any kind, neither the desired, nor the undesired can cause one to waver. Steadfast is his mind, gained is his deliverance. (Anguttara Nikaya VI, 55)

Buddhist scholars and others have noted that the monk-arhat could be thought of as selfish because of his concentration on his own salvation and not that of others.[179]

The creators of this manuscript emphasized the karmic and didactic aspects of Buddhism by the inclusion of personages important to Buddha, and to the growth and development of Buddhism. This was deemed important for the continuation of the linkages to the Buddha's thoughts and the Dhamma. The pages of the manuscript devoted to the arhats, religious male and female laity constitutes 24 percent of the pages of the manuscript. Ten women of great importance are included along with 103 men. Males included members of the Sangha, distinguished lay and religiously eminent people. As can be seen in the following sketch, 11 of 45 pages are devoted to depictions of arhats and religious personages. The grouping of manuscript pages containing the arhats and religious personages is concise and orderly.

The presentation of the people important to Buddha and Buddhism follows the unfolding of their appearance in the manuscript; four categories are presented in the following order:

The Monks-Arhats – 80 persons

Male Religious Personages Accorded Pre-Eminence[LXIII] – 13 persons

Male Laity Personages Accorded Pre-Eminence – 10 persons

Female Personages – 10 persons

Besides the name, the description presented for monks in the manuscript is brief, and mainly consists of one-line sentences. It may be speculated that the monk, or learned lay person presenting the manuscript to a group of laity, or novice monks, may have embellished these descriptions, or in case of a name only, perhaps have added his own knowledge about the person shown.

When discovered, amplification for the reasons of importance of the arhat are provided *in italics*.

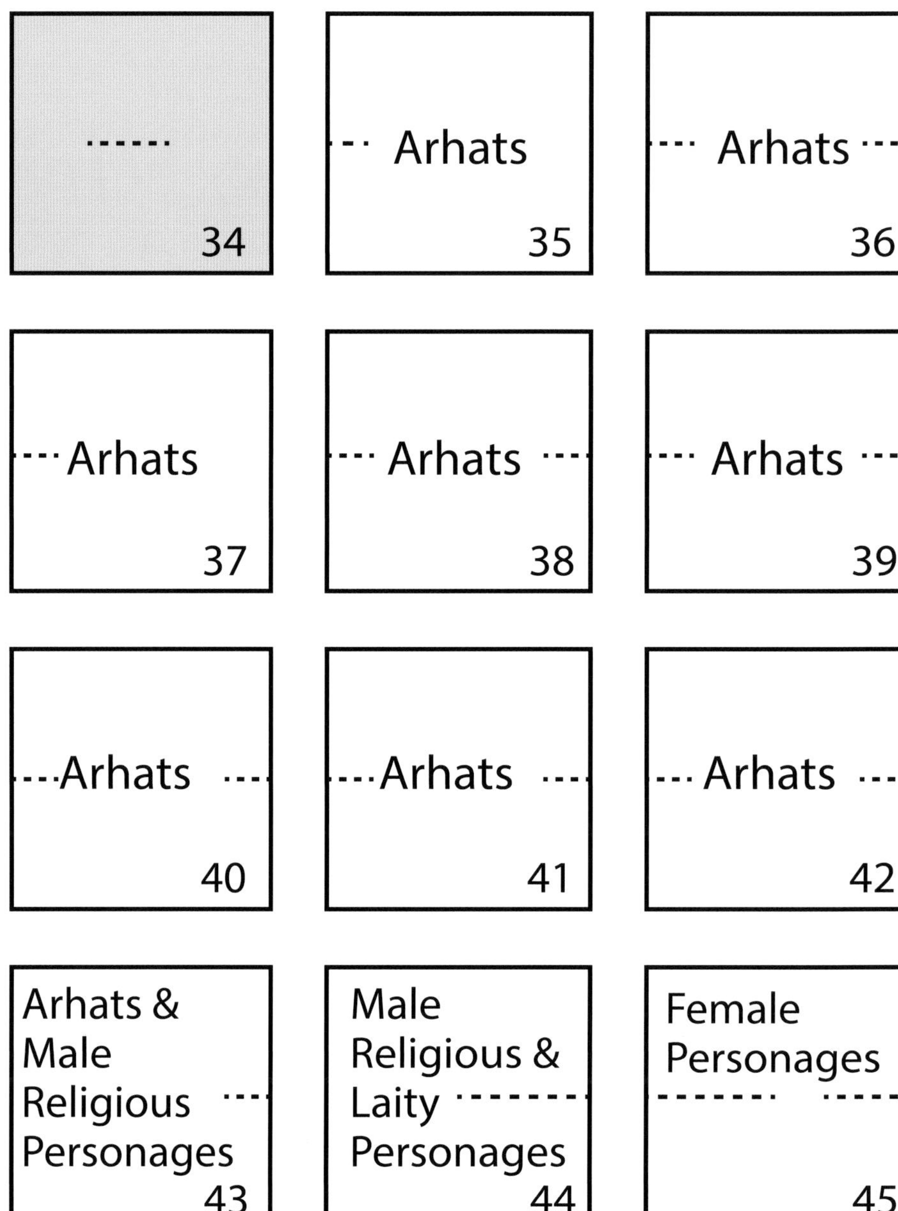

LXIII An award, or recognition not clarified.

Arhats and Their Importance to Buddhism

The 103 male arhats, religious and laity personages contained and illustrated in the manuscript require identification greater than either a name, or a brief one line synopsis provided in the manuscript, to fulfill the status they achieved and their roles in the development of Buddhism as a world force. One way to do this is to consult what others have thought and written. Over the last two and half millennium there have been listings developed of monks and arhats who were thought to be crucial to the development of Buddhism. Three influential known listings are:

1. The ten most important disciples of the Buddha who are most frequently mentioned in the sūtras. This small grouping of the ten most important Buddhist saints is undisputable and to students of Buddhism is a 'no brainer'. This list is the basic grouping of the 'super' arhats.
2. The listing sixteen or eighteen arhats initially charged with the continuation and protection of Buddha's teachings. This listing of Arhats was developed over the centuries in Korea, China, Japan and various other seats of Buddhism.
3. The ninety-three eminent disciples of the Buddha. Developed from a treatise by the Venerable Piyaratana Māhathera of Ceylon which is the basis for C. De Saram's book. *Pen Portraits Ninety Three Eminent Disciples of the Buddha*. The listing by a Ceylonese Theravāda scholar of monks and *arhats* who were ranked in their importance to Buddha bears significance to this Burmese Theravāda manuscript which presents a sizable grouping of arhats for the consideration of the reader.

In the following tables of listings the Arhats identified in the manuscript are presented with bold type along with their manuscript page number.

The 10 Great Disciples of the Buddha

The ten most important companions of the historical Buddha are mentioned in the Mahayana sutras, Sanskrit, Pali, and other texts. The ten listed not only lived with the Buddha but, with time in the passage of Buddhism, were worshipped in their own right an action currently no longer practiced. The 10 disciples of the Buddha are shown in Table 7. It has been noted that 'different texts place them in various orders of priority'.[180]

TABLE 7. THE 10 GREAT DISCIPLES OF THE BUDDHA

Rank	Name	Manuscript page number
1	Maha Kāśyapa	35
2	Anandā	40
3	Sāriputta	35
4	Subuti	42
5	Puṇṇa	41
6	Moggallana	35
7	Kinkhaā Revata	40
8	Aniruddha	40
9	Upāli	40
10	Rāhula	41

The 16 Arhats

The groupings of arhats important to Buddha and Buddhism range in number from 10 to upwards of 500, or even multiples of this figure, who accompanied Buddha on his journeys; the figure can be taken just to mean a large crowd.[181] Important is the fact that only sixteen or eighteen of the many of Arhats are actually worshipped.[182]

An apocryphal legend from the *Ekottara-agama Sutra* (a discourse on the Buddha translated from Sanskrit into Chinese 313 C.E.) tells of Buddha when Buddha became aware that death was near he appointed four arhats to protect his teachings. They were Maha Kāśyapa, Piṇḍola Bhāradvāja, Kundashana and Rāhula. They were ordered to remain in existence to protect his teachings until the next Buddha arrived.[183]

Later in 437 or 439 C.E. a Buddhist Chinese monk translated the Mahayana Buddhist version of *Ekottara-agama Sutra* wherein the four arhats became sixteen and then later eighteen. The legend of the sixteen or eighteen arhats grew over the centuries and the legend moved from China to Korea and then to Japan.[184]

The following tabulation of the sixteen arhats is from a book by Reginald A Ray, *Buddhist Saints in India: A study in Values & Orientations* (Oxford University Press, New York, 1999). Ray's book is focused on Buddhist saints. Two other listings are shown in Appendix C. There is a commonality, in that the first six rankings of the other two are in total agreement with Ray's. Table 8 is a listing of the 16 *arhats* with those portrayed and profiled in the manuscript shown in bold along with their manuscript page numbers.

TABLE 8.
A RANKING AND INCLUSION LISTING OF THE 16 ARHATS ACCORDING TO REGINALD A. RAY[185]

Rank	Name (honorific)	Manuscript page number
1	**Piṇḍola Bhāradvāja**	39
2	Kanaka Vatsa	
3	Kanaka Bharadvaja	
4	**Subuti**	42
5	**Nālaka**	36
6	**Baddiya**	35
7	**Kāludāyi**	35
8	Vajraputra	
9	Supaka	
10	**Payantakethe**	40
11	**Rāhula**	41
12	Nagasena	
13	Ingada	
14	Vanavasa	
15	Ajita	
16	Chudapanthaka	

The 93 Eminent Disciples of Buddha

Table 9 lists the monks and *arhats* who were ranked in their importance to Buddhism. This was developed from a treatise by the Venerable Piyaratana Māhathera[LXIV] of Ceylon (Sri Lanka) which was translated, and authored, by C. De Saram in a book *Pen Portraits Ninety Three Eminent Disciples of the Buddha.* Not all arhats in the manuscript have been assigned a rank of importance to Buddha and Buddhism; some may have not been identified for reasons of nomenclature, translation and differences in identification between Burma and Sri Lanka. However, it is notable that the discussion in this manuscript opens with the first three arhats Kondańña, Sāriputra, and Moggallana which are also ranked the first three of importance to Buddhism in C. De Saram's book? The *arhats* portrayed and profiled in the manuscript are shown in bold along with manuscript page numbers.

LXIV Details, dates and other information of the Venerable Piyaratana Māhathera of Ceylon have been sought, but yet have to be found.

TABLE 9. EIGHTY MALE ARHAT PERSONAGES Listed by Venerable Piyaratana Māhathera of Ceylon in Rank of Importance to Buddhism

Rank	Name (Burmese honorific)	Manuscript Page Number	Rank	Name (Burmese honorific)	Manuscript Page Number
1	**Kondańńa**	35	41	Mogharajā	
2	**Sāriputta,**	35	42	Wappa≠	
3	**Moggallana**	35	43	**Baddiya**	41
4	**Maha Kāśyapa (Shin)**	38	44	**Māhanan (Shin)**	36
5	**Aniruddha**	40	45	**Assaji**	36
6	**Baddiya**	35	46	**Kimbila**	41
7	**Lakuddaka Bhaddiya**	39	47	Bhagu	
8	**Piṇḍolabhāradvāja (Shin)**	39	48	Cuna	
9	**Puṇṇa**	41	49	**Nālaka**	36
10	Māha Kaccāna		50	Seludaiyi	
11	Cula Panthaka		51	Bhaddali	
12	Maha Pańthaka		52	Ajita	
13	Subuti	42	53	Tissa Metteyya	
14	**Khadiravaniya Revata**		54	Punnaka	
15	Kinkhaā Revata	40	55	Mettagu	
16	Kolivisa Sona		56	Dhotaka	
17	Khutikanna Sona		57	Upasiwa	41
18	**Sivali**	41	58	Nandaka	
20	**Rāhula**	41	59	Hēmaka	
21	**Rahttapāla**	37	60	Thodeyya	
22	**Kondadāna**	37	61	Kappa	
23	**Vigisa**	37	62	Jatū-Kańńi	
24	Wanganta-Putta Upasena		63	Bhadravudha	
25	**Dabba**	41	64	Udaya	
26	**Pilinda Vaccha**	36	65	Posala	
28	**Kumāra Kassapa**	36	66	Pingiya	
29	**Konhitita (Shin)**	40	67	**Meghiya**	38
30	**Anandā**	40	68	Chńńaa	
31	**Ūruvela Kassapa**	38	69	Upavana	38
32	**Kāludāyi**	35	70	Bhaddaji	
33	**BāKula**	40	71	**Yasa**	36
34	**Sobhita**	36	72	Lakkhana	
35	**Upāli**	40	73	Gavampati	
36	Nanddaka		74	**Punnaji (Shin)**	37
37	**Nanda**	40	75	**Nadi Misspa**	38
38	Māha Kappina		76	**Ańgulimalā**	35
39	**Sāgata**	38	79	**Sabhiya**	37
40	Rāddha		80	Sakuludāyi	

In the case of the eminent female personages only one has been reconciled with the brief description given in the manuscript and the De Saram's description.

TABLE 10. THIRTEEN EMINENT FEMALE PERSONAGES
Listed by Venerable Piyaratana Māhathera of Ceylon in Rank of Importance to Buddhism

Rank	Name	Manuscript page Number	Rank	Name	Manuscript Page Number
1	**Dukula Mater** (Maha Pajāpati Gotami)	45	8	Sakulā	
2	Khēma		9	Bhadda-Kundali-Kesa	
3	Uppalavaṇṇā		10	Badda-Kapilāna	
4	Patācāra		11	Badda-Kaċċana	
5	Dhammina-Dinnā		12	Kisa Gotami	
6	Sundari Nanda		13	Sigalaka Māta	
7	Sonā				

Paintings of Arhats, Monks and Laity

In the Burmese manuscript the interaction in the paintings between the arhats and that which occurs in the depiction of the women (shown last) differs. The *arhats* are separated from one another by imaginative flowers on thin stems, and focus their gaze towards the center of the picture and are seated on multi-colored plinths, or mats. An iconographical conformity is that all fingers and toes are of the same size. This apparent anomaly is in conformity to Burmese and Thai iconography. The posture of the arhats is either shown with an open right or left hand touching the chest and the palm of the other hand on the opposite knee. This posture represents one of strong resolve on the part of that monk, to enter Nibbana.

The monks who constituted the early religious aristocracy of Buddhism are shown in this manuscript as ethereal young men, even though most of the arhats depicted had lived to a very old age (see following text box which reinforces this thought).

Since this is a Burmese manuscript, the illustrations contain cultural characteristics of the time it was written.

- The honorific 'Shin' and 'Ashin', Burmese for 'Wise Monk', is used for a number of these personages but not all.
- Dress and costume of the laity depicted are Burmese and of the period.

Appendix D is an alphabetical listing of the monks/arhats depicted. The list includes rankings of importance to the Buddha and Buddhism.

STATUES AND WAX IMAGES OF DEPARTED MONKS

Thai monks, who have devoted a lifetime to the Sangha, are honored and remembered in a number of ways after their death. In Thai temples there are life-like wax images of monks who have meant much to their communities. These statues, either in wax or bronze, all depict serene aged monks without the intrusion of any younger monks into the groupings.

At Wat Sang Kaew, Mae Suai, Chiang Rai, Thailand, a number of bronze statues of departed monks sit in honor on low tables.

At Wat Phra Singh, Chiang Mai, Thailand life-like wax images of departed monks have been placed in honor in the main wihan.

Explanation of the Captions

Kondañña: Name from common Buddhist literature usage; exception if monk is not known, only identified, then as transliterated

Kondañña: 'Who knows the best time; who was first enlightened.' (T) = Translation from manuscript

Kondañña was one of the five ascetics Buddha revisited after Enlightenment. The Buddha preached the Four Noble Truths and the Middle way to them, which Kondañña alone understood at once. The Buddha called out "Kondañña knows" and made him the first Buddhist monk. – Referrals to other sources have been made to enhance knowledge of the arhat shown. This abbreviated information is shown in italics.

Kondañña is ranked 1st in importance to the Buddha and Buddhism, following the enlightened One himself. Sourced from C. De Saram's book *Pen Portraits Ninety Three Eminent Disciples of the Buddha.*

❊ ❊ ❊

The illustrations of personages begin with the three most illustrious of the Buddha's disciples: Kondańńa, Moggallana and Sariputta. Ańgulimalā, the fourth illustrated, arrived into the Sangha from a murderous past life.

Top row: Kondańńa, Ashin Sariputta, Ashin Moggallana, Angulimala

Bottom row: Baddiya, Vappa Htera, Vakkali, Kaludayi

Manuscript Page 35

The Arhats and their associated translations are shown left to right, then top to bottom.

Kondańńa: 'Who knows the best time; who was first enlightened.' (T)

Kondańńa was one of the five ascetics Buddha revisited after Enlightenment. The Buddha preached the Four Noble Truths and the Middle way to them, which Kondańńa alone understood at once. The Buddha called out "Kondańńa knows" and made him the first Buddhist monk.

Kondańńa is ranked 1st in importance to the Buddha and Buddhism, following the enlightened One himself.

Sāriputra: 'The most educated person after the Buddha.'(T)

Sāriputta was acknowledged by the Buddha as the foremost of all disciples in the attainment of wisdom. Born a Brahmin, he became a Buddhist with his friend Moggallana after hearing a discourse by one of the first five disciples of Buddha. He was the Buddha's first chief disciple on the right-hand side.

Sāriputta is ranked 2nd in importance to Buddhism.

Moggallana: The most powerful supernatural power after the Buddha.' (T)

Moggallana was acknowledged by the Buddha to have great attainment of miraculous powers; he became the Buddha's second main disciple on his left. He was foremost in the propagation of the faith, blessed with great psychic powers, and able to convey messages from heaven and hell.

Moggallana is ranked 3rd in importance to the Buddha and Buddhism.

Aṅgulimalā: Name only transliterated.

Aṅgulimalā, dubbed 'a man of extremes'[186] *is most interesting; he was a former dacoit who converted after a famous meeting with the Buddha. His name means 'Garland of Fingers' from his practice of cutting a finger off all his murdered victims. He became one of the 80 great arhats of the Noble Order.*

Aṅgulimalā is ranked 76th in importance to Buddhism[LXV]

Baddiya: 'The highest lineage.' (T)

One of the arhats entrusted by the Buddha with the Dhamma.[187] *The son of the King of Kaligoda, a Sakyan, Baddiya ranked the highest in birth status of all the arhats. The father of the Buddha urged the Sakyans to meet the Buddha and Baddiya was part of that group; as a king, he became a member of the Sangha.*[188]

Baddiya is ranked 6th in importance to Buddhism.

Vakkali: 'The most devoted person to the Buddha.' (T)

Vappa Htera: Name only transliterated.

Kāludāyi: 'The most clever person; devoted to the relatives.' (T)

Kāludāyi was one of the arhats entrusted by the Buddha with the Dhamma.[189] *He was the son of a minister to the Buddha's father and was born on the same day as the Buddha. Because of this notable event Kāludāyi became one of the Buddha's birth treasures.*[190]

Kāludāyi is ranked 32nd in importance Buddhism.

Manuscript Page 36

Assaji: Name only given in the manuscript.
The chief disciple of Sariputta.[191]

Assaji is ranked 45th in importance to Buddhism.

Shin Māhanan: Name only given in the manuscript.
He was one of the first five arhats who entered the Noble Order.[192]

Shin Māhanan is ranked 44th in importance to Buddhism.

Saha Udāyi Htera: Name only transliterated.

LXV The section devoted to Aṅgulimalā in C. De Saram's book, ***Pen Portraits: Ninety Three Eminent Disciples of the Buddha,*** is almost equal in length to the total of the first three *arhats* of importance to Buddhism.

Top row: Assaji, Shin Māhanan, Saha Udāyi Htera, Pilinda Vaccha

Bottom row: Yasa, Nālaka, Sobhita, Kumāra Kassapa

Pilinda Vaccha: 'The most devoted of the gods.' (T)

The Buddha when addressing the Noble Order and the laity declared that among the arhats, Pilinda Vaccha was most pleasing to the gods.[193]

Pilinda Vaccha is ranked 26th in importance to Buddhism.

Yasa: Name only transliterated.

Yasa, a disciple of Anandā, had a pivotal role in the Council of Vaiśālī, correcting the illicit practices of the Vaijian monks.[194]

Yasa is ranked 71st in importance to Buddhism.

Nālaka: Name only transliterated.

Nālaka was one of the arhats entrusted by the Buddha with the Dhamma.[195]

Nālaka is ranked 49th in importance to Buddhism.

Sobhita: 'The best person to know past lives.' (T)

'Sobhita was able to recall his past lives in order of succession for a period of 500 kalpas.'[196] *'He was said to be an exponent of Abhidhamma.'*[197]

Sobhita is ranked 34th in importance to Buddhism.

Kumāra Kassapa: 'The best speaker of Dhamma.' (T)

Kumāra Kassapa was an elder renowned for his eloquent and brilliant imagery while preaching sermons. He was ordained when 20 instead of 21 because the Buddha reasoned that as his mother was a nun when he was born, his time in the womb counted in the reckoning of his age.[198]

Kumāra Kassapa is ranked 28th in importance to Buddhism.

Top row: Vimala Htera, Subāhu Htera, Rahttapāla, Vigisa

Bottom row: Shin Punnaji, Shin Tira, Sabhiya, Shin Negathen

Manuscript Page 37

Vimala Htera: Name only transliterated.

Subāhu Htera: Name only transliterated.

Rahttapāla: 'The most generous person to become a monk.' (T)

The only son of the king of Kurus, he converted after listening to a sermon by Buddha and asked to be ordained. He was refused ordination as he was underage; additionally, as the only heir to the kingdom, his parents strongly objected. Rahttapāla remained adamant and was accepted into the Sangha but had to wait 12 years to become an arhat. The Buddha restricted him from visiting his parents during his probation. Rahttapāla were cited by the Buddha in an address to the monks for having renounced his riches and the world for reasons of faith.[199]

Rahttapāla is ranked 21st in importance to Buddhism.

Vigisa: 'The best poem writer.' (T)

Vigisa was a born poet. He recited his poems to the Buddha. The Buddha while

addressing the monks said that Arhat Vigisa always had the correct word when speaking to suit the occasion.[200]

Vigisa is ranked 23rd in importance to the Buddha and Buddhism.

Shin Punnaji: Name only given in the manuscript.

The son of the treasurer of Benāres was one of the four friends of Yasa to join the Noble Order.[201]

Shin Punnaji is ranked 74th in importance to Buddhism.

Shin Tira: Name only transliterated.

Sabhiya: Name only given in the manuscript.

Sabhiya, a powerful speaker, skilled debater, and teacher still sought knowledge. After a series of disappointments with various teachers he found the Buddha and, falling at Buddha's feet requested ordination and became an arhat.[202]

Sabhiya is ranked 79th in importance to Buddhism.

Shin Negathen: Name only transliterated.

Manuscript Page 38

Nadi Kassapa: Name only given in the manuscript.

Nadi Kassapa was the younger brother of Ūruvela Kassapa. Nadi had 300 followers and Ūruvela 500. The Buddha converted Ūruvela and Nadi and their followers, after a sermon by the Buddha called 'Adittha Pariyay Sutta'. Nadi was one of the 80 great arhats of the Noble Order.[203]

Nadi Kassapa is ranked 75th in importance to Buddhism.

Ūruvela Kassapa: 'The monk with the largest audience.' (T)

Before Ūruvela Kassapa met the Buddha, Ūruvela Kassapa had a large number of followers that only the Buddha himself could convert such a prominent and strong leader. Buddha declared, when addressing the Sangha, that Ūruvela was pre-eminent among the great Arhats.[204]

Ūruvela Kassapa is ranked 31st in importance Buddhism.

Top row: Nadi Kassapa, Ūruvela Kassapa, Upavana, Meghiya

Bottom row: Shin Maha Kāśyapa, Shin Yawneyhtey, Sāgata, Shin Nāgita

Upavana: Name only given in the manuscript.

Upavana, well versed in Vedic lore, was an attendant of the Buddha before Anandā assumed this task. He attended the Buddha in his last illness and he was one of the 80 great Arhats of the Noble Order.[205]

Upavana is ranked 69th in importance to the Buddha and Buddhism.

Meghiya: Name only given in the manuscript.

Meghiya, who hailed from the Sakyan royal family, was one of the attendants of the Buddha until relieved by the Buddha's cousin Anandā. Meghiya also was one of the 80 great Arhats of the Noble Order.[206]

Meghiya is ranked 67th in importance to Buddhism.

Shin Maha Kāśyapa: 'The best practitioner of the 13 Dhutanga practices in the

Theravāda tradition that mark the limits of monks who wish to pursue asceticism to attain Nirvana.' (T)[207]

Kassapa (Pali) or Kāśyapa (Sanskrit) was, at first, one of Buddha's ten great disciples. Later, after the death of the Buddha, he took over Sangha. leadership of the His great strengths were his ascetic self-discipline and moral strictness. He convened the first Buddhist council which addressed concerns of emerging trends towards superficiality and less strict lifestyles among members of the Sangha. He is also known as Maha Kāśyapa, i.e. the great Kāśyapa.[208]

Shin Maha Kāśyapa is ranked 4th in importance to Buddhism.

Shin Yawneyhtey: Name only transliterated.

Sāgata: 'The best observer of fire; concentration on fire and flames to achieve tranquility.'

Sāgata was cited by the Buddha for his ability to enter into a trance.[209]

Sāgata is ranked 39th in importance to Buddhism.

Shin Nāgita: Name only transliterated.

Top row: Shin Mhakappina, Shin Mahakissayana, Lakuddaka Bhaddiya, Shin Piṇḍolabhāradvāja

Bottom row: Shin Mahāsudda, Shin Konhtita, Shin Maha Yakalocaweto, Sulapanna

Manuscript Page 39

Shin Mhakappina: 'The best teacher of 1,000 monks.' (T)

Shin Mahakissayana: 'The best giving short informative remarks but with the knowledge and skill leading to greater depth.' (T)

Lakuddaka Bhaddiya: 'The best voice.' (T)

Lakuddaka Bhaddiya was a small man with a melodious voice. His melodious voice was reported surpassed only by the Buddha's voice. The Buddha declared that Lakuddaka Bhaddiya was pre-eminent in the Noble Order for his voice.[210]

Lakuddaka Baddiya is ranked 7th in importance to Buddhism.

Shin Piṇḍolabhāradvāja: 'The brave voice like a lion.' (T)

Shin Piṇḍolabhāradvāja joined the Sangha because of greed; he liked to eat and had a large alms bowl which he kept under his bed at night. Over the course of time as a monk he overcame his appetites and was held in esteem and came to be regarded as an arhat. He was recognized as a master and a forest saint whose path was defined by meditation.

Shin Piṇḍolabhāradvāja is ranked 8th in importance to Buddhism

Shin Mahāsudda: Name only transliterated.

Shin Konhtita: 'The monk who has the best enlightenment.' (T)

Shin Konhtita was well versed in the four attainments for which Buddha cited him. The Pāli words are (1) Attha (2) Dhamma (3) Nirutti and (4) Patibbāna."[211]

Shin Konhtita is ranked 29th in importance to Buddhism

Shin Maha Yakalocaweto: Name only transliterated.

Sulapanna: 'The best person for creating several persons in mind.' (T)

Manuscript Page 40

The Arhats and their associated translations are shown left to right, then top to middle tier and then to bottom tier.

Top row: Upāli, Aniruddha, Bākula, Yasodha

Middle row: Ānandā, Kinkhaā Revata, Nanda, Dārusi

Bottom row: Shin Payantakethe, Shin Sannakha Htay Meinda, Shin Bawe Tohtay, Shin Taktha Sitay Thij Htay

Upāli: 'The best learner of disciplines of monks.' (T)

Upāli, a barber by profession, was ordained before six royal princes who accompanied him. Because of this precedence Upāli, was given precedence in the order of monks. 'This confirmed the Buddha's desire to ignore caste distinctions. Upāli learned the Vinaya Pitaka by heart and became an authority on discipline. He was so acknowledged by the Buddha.[212]

Upāli is ranked 35th in importance to the Buddha and Buddhism.

Aniruddha: 'The one gifted with the best Dibbacakku.'[LXVI] (T)

Aniruddha, one of ten of the Buddha's great disciples whose name is constantly mentioned in the sūtras.

Aniruddha is ranked 5th in importance to Buddhism.

Bākula: 'The best health.' (T)

In a time of low life expectancy Bākula was noted for his freedom from disease. Bākula a Brahmin, who was learned in Vedic lore, became dissatisfied with his life and became a hermit in the Himalayan Mountains. Remarkably free of disease in his 80th year, he heard the Buddha and was ordained. The Buddha addressing a group of monks noted that Bākula was the most healthy and free of all disease.[213]

Bākula is ranked 33rd in importance to the Buddha and Buddhism.

Yasodha: Name only transliterated.

Ānandā: 'The best memory to learn the Buddha's teaching.' (T)

Ānandā was Buddha's cousin and one of his main disciples. Because of Ānandā's intercessions with the Buddha, women were allowed to become nuns. He also played a major role in the 'First Buddhist Council.

Ānandā is ranked 30th in importance to Buddhism.

Kinkhaā Revata: 'The one with the best supernatural power.' (T)

Kinkhaā Revata enjoyed going into trances, sometimes for seven days at a time, and the bliss that afterwards accompanied them. He loved abodes of great solitude and the delights that sprung from meditation.

Kinkhaā Revata is ranked 15th in importance to Buddhism.

Nanda: 'The best observer of the five senses.' (T)

Nanda is believed to be one of the arhats who remained in the world as protectors of the dharma.[214]

Nanda is ranked 37th in importance to Buddhism.

LXVI Dibbacakku = divine eye; i.e. clairvoyance, supernatural powers to see future happenings.

Dārusi: 'The one who learnt the Dhamma fastest.' (T)

Shin Payantakethe: Name only transliterated.

Payantaketh was one of the arhats entrusted by Buddha with the Dharma.[215]

Shin Sannakha Htay Meinda: Name only transliterated.

Shin Bawe Tohtay: Name only transliterated.

Shin Taktha Sitay Thij Htay: Name only transliterated.

Top row: Baddiya, Kimbila, Sona, Upasiwa

Middle row: Sivali, Shin Mathay, Puṇṇa, Shin Mata Buhtay

Bottom row: Dabba, Rāhula, Kondadāna, Ashin Tawka Thuma Hthy

Manuscript Page 41

Baddiya: Name only given in the manuscript.

Baddiya was one of the five mendicants who shared companionship and privation with Siddhartha before he attained Enlightenment and discovered the Middle Way. When the Buddha delivered the first sermon, only Kondańńa absorbed and understood it fully. Buddha, gradually over a period of time instructed Baddiya and the other three companions and finally, at the next sermon, they all understood and became arhats.

Baddiya is ranked 43rd in importance to Buddhism.

Kimbila: Name only given in the manuscript.

Kimbila was one of the aristocrats who accompanied Upāli, the barber, to meet the Buddha and become arhats.

Kimbila is ranked 46th in importance to Buddhism

Sona: Name only transliterated.

Upasiwa: Name only given in the manuscript.

Upasiwa asked the Buddha a fundamental question of life after death at a gathering of leaders. Buddha responded that the termination of life no matter how long must inevitably be unsatisfactory if Nibbana had not been realized. Upasiwa and his followers understood and became arhats.[216]

Upasiwa is ranked 57th in importance to Buddhism.

Sivali: 'The one who was foremost among those who received offerings.' (T)

Sivali was the foremost in the receipts of offerings and accompanied the Buddha on arduous journeys.

Sivali is ranked 18th in importance to the Buddha and Buddhism.

Shin Mathay: This may not be a name; Shin' and 'Mathay' are both honorifics given to Burmese members of the Sangha. Thus, this illustration may simply be 'a space filler'.

Puṇṇa: Name only given in the manuscript.

Puṇṇa, a forest monk was identified by a group of his peers as excelling in matters of poverty, contentment, seclusion, and meditation.[217]

Puṇṇa is ranked 9th in importance to the Buddha and Buddhism.

Shin Mata Buhtay: Name only transliterated.

Dabba: 'The best organizer of beds for the monks.' (T)

Dabba became renowned for his abilities to prepare for visiting monks whom the Buddha recognized as important[218]

Dabba is ranked 25th in importance to the Buddha and Buddhism.

Rāhula: 'The best keeper of disciplines.' (T)

Rāhula was the Buddha's son. He entered the Sangha as a child of seven and thus is considered the guardian of novices. He was ordained into the Sangha by Sariputta (Folio 1.) and probably died young before his father. Rāhula is among the ten great disciples of the Buddha.

Rāhula is ranked 20th in importance to the Buddha and Buddhism.

Kondadāna: 'The one who has the best of luck.' (T)

Kondadāna was cited by the Buddha for invitations to receive alms.[219]

Kondadāna is ranked 22nd in importance to the Buddha and Buddhism.

Ashin Tawka Thuma Hthy: Name only transliterated.

Manuscript Page 42

Subuti: 'The best dweller from non-attachments; the best person to receive donations.' (T)

Subuti was deemed to be faultless, without meanness whatsoever. His preaching was straightforward; strong and deemed to be Buddha-like. The Buddha cited Subuti as free from moral depravity.[220]

Subuti is ranked 13th in importance to the Buddha and Buddhism.

Upasena: 'The most glorious person.' (T)

Sona: Name only transliterated.

Ashin Taytahta Htay: Name only transliterated.

Revata Pandit: 'The person who likes living in the forest the best.' (T)

Ashin Kumara: 'The person who learns quickly.' (T)

Ashin Zartura Pandit: Name only transliterated.

Ashin Thu Nad: Name only transliterated.

Ashin Tanwara: Name only transliterated.

Ashin Taung Phela: Name only transliterated.

Unknown Arhat: No script provided; illustration may have been added for compositional balance.

Ashin Take Tah: Name only transliterated.

Top row: Subuti, Upasena, Sona, Ashin Taytahta Htay

Middle row: Revata Pandit, Ashin Kumara, Ashin Zartura Pandit, Ashin Thu Nad

Bottom row: Ashin Tanwara, Ashin Taung Phela, Unknown Arhat, *Ashin Take Tah*

Top row: Shin Kotikanda Thawana, Shin Mandanee Potta Ponna, Shin LiPaw The Le, Shin Mawgara Raza

Bottom row: Sona Hteiri Nidaniana, Nardra Hteiri, Keima Hteiri, Ou Paiawon Hteiri, Padasari, Dhamma Nidaninna Hteiri

Manuscript Page 43: Arhats and Male Religious Personages Accorded Pre-Eminence

This page begins with the continuation of the arhats; all of the four arhats are in the posture of resolution to enter Nibbana. The lower part of the page has seven arhats who have been accorded pre-eminence. The posture of this group varies with four in resolution to enter Nibbana and the other in various unknown postures. They are seated on simple cushions, or thrones, instead of flower-type plinths. These monks are separated by buds of red and white lotuses.

Arhats

Shin Kotikanda Thawana: 'The person whose talent lies in reading of the Dhamma.' (T)

Shin Mandanee Potta Ponna: 'A good lecturer of the Dhamma.' (T)

Shin LiPaw The Le: Name only transliterated.

Shin Mawgara Raza: 'The saver of old rope.' (T)

Male Religious Personages Accorded Pre-Eminence

Sona Hteiri Nidaniana: 'Accorded pre-eminence in demonstrating great diligence in his work.' (T)

Nardra Hteiri: 'Accorded pre-eminence for his ability to go into a deep state of concentration.' (T)

Pahta Patigo Tami Hteiri: 'Accorded pre-eminence in his abilities to distinguish between day and night.' (T)

Keima Hteiri: 'Accorded pre-eminence for his wisdom.' (T)

Ou Paiawon Hteiri: 'Accorded pre-eminence in potency.' (T)[LXVII]

Padasari: 'Accorded pre-eminence because of his ability to strictly adhere to the Vinayas, the monk's code of conduct.' (T)

Dhamma Nidaninna Hteiri: 'Accorded pre-eminence in teaching morals based on the Dhamma.' (T)

Manuscript Page 44: Male Religious and Male Laity Personages Accorded Pre-Eminence

The monks lack insertions of flowers between them. They all have names which end the same: 'Hteiri'; the meaning, if an honoree, or family connection, has not been uncovered. The male laity appears to be interacting between themselves. The dress commonality between the men are the longyi (sarong) and head dresses. Some of the men are wearing collarless jackets others cloth wraps. The differing dress may represent ethnic differences, i.e. Shan, Mon, Kaya etc.

Arhats

Peingala Hteiri: 'Accorded pre-eminence in generosity.' (T)

Butta Kisana Hteiri: 'Accorded pre-eminence in intellectual ability (*Jathoidaja*).' (T)[221]

Thaku La Hteiri: 'Accorded pre-eminence in awareness of superhuman vision.' (T)

Buddha Kuda La Kathi Hteiri: 'Accorded pre-eminence in awareness of the Dhamma.' (T)

Buddha Kapilani Hteiri: 'Accorded pre-eminence in awareness of *Pubbeiniwatha nouthati*[222] Dhamma.' (T)

Kithagawdami Hteiri: 'Accorded pre-eminence in the folding of robes.' (T)

Male Laity Personages Accorded Pre-Eminence

Hahta Lawaka: 'Accorded pre-eminence for providing an audience with the four rules of good social behavior.' (T)

Hahtigamanel: 'Accorded pre-eminence in the serving of the Sangha.' (T)

LXVII Translation confirmed.

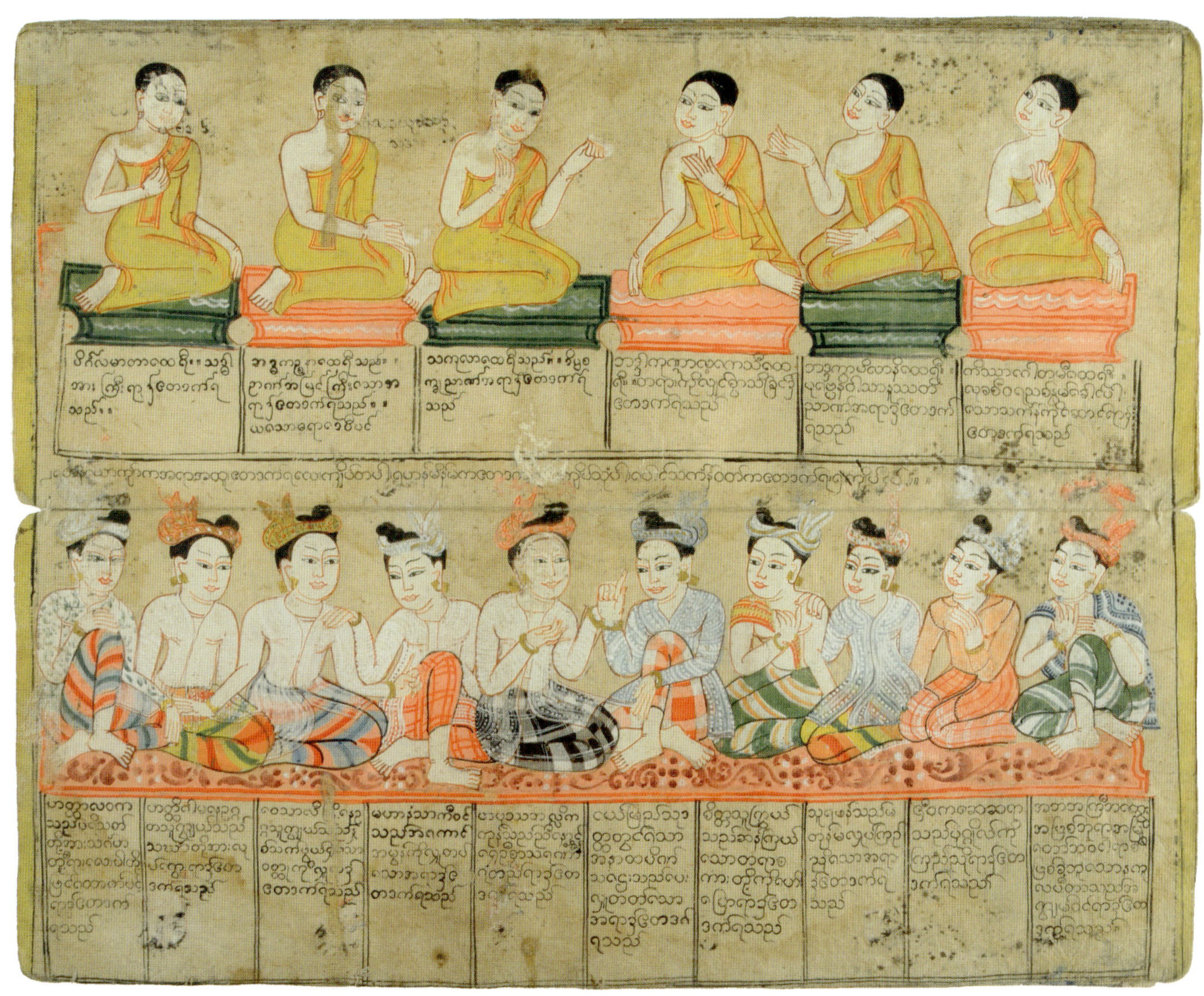

Top row: Peingala Hteiri, Butta Kisana Hteiri, Thaku La Hteiri, Buddha Kuda La Kathi Hteiri, Buddha Kapilani Hteiri, Kithagawdami Hteiri

Bottom row: Hahta Lawaka, Hahti-gamanel, Eggam, Mahanan, Ta Hpou Tha, Anatapine, Seita, Thura Pan, Ziwaka, Nakua La Pita

Eggam: 'Who was rich, was Accorded pre-eminence in the offering of nice items.' (T)

Mahanan: A member of the Thaki generation[223] was Accorded pre-eminence in the offering of good items.' (T)

Ta Hpou Tha: 'Ta Hpou Tha and his brother Balika were Accorded pre-eminence for having a firm belief in the Triple Gems.' (T)

Anatapine: 'Who was rich and named a Thuddahta[224] when he was young was Accorded pre-eminence for making donations to charity.' (T)

Seita: 'A rich man, was Accorded pre-eminence for the preaching of a wonderful sermon.' (T)

Thura Pan: 'Accorded pre-eminence for having a deep reverence for the Buddha.' (T)

Ziwaka: 'A herbalist, was Accorded pre-eminence for having personal reverence.' (T)

Nakua La Pita: 'Accorded pre-eminence as someone who had experienced five hundred lives. He was the grandfather of the Buddha and familiarizing himself with the Dhamma.' (T)

Left to right: Dukula Mater, Suppiya, Sāmāwati, Khujjuttara, Visākhā, Sujātā, Uttarā, Suppavāsā, Kātiyāni, Kali

Manuscript Page 45: Notable Female Laity Personages

The female personages are pictured as a single group of women, like the male laity, appears to be interacting among themselves. They appear to be enjoying the togetherness of the moment. The women are shown as mature, and young women attired in Burmese dress of the period sharing cushions.

The following are translations from the manuscript of the women who were praised by the Buddha as noblest persons for various reasons.

Dukula Mater had the best friendship with Buddha and his Sangha. She was a mother, a step mother, or an aunt, 500 times during each of the Buddha's previous lives. (T)

Suppiya, from Baharnasi (Banares), provided the best treatment for sick monks. (T)

Sāmāwati, a queen, was a person filled with metta, loving-kindness. (T)

Khujjuttara, servant of Queen Sāmāwati, one the most educated persons and blessed with good general knowledge. (T)

Visākhā the most generous person for donations. (T)

Sujātā, the first person who took refuge in the Buddha and Dhamma before the Sangha started. (T)

Uttarā possessed the most concentrated focus on *Metta,* of kindness and love. (T)

Suppavāsā, the best person who donated the best things. (T)

Kātiyāni, the most devoted person to the Buddha. (T)

Kali, the most devoted person to the Buddha who only heard his name. (T)

PART 6

Selected Heavenly Texts

Inquiry into the script associated with the illustrations of the Theravāda heavens proved interesting. Only a sampling of the script was translated to develop a flavor of and an indication of what appeared to be important to the donor and scribes.[LXVIII] The translations may appear to be extensions of King Lithai's book on mind types and rebirth possibilities of the inhabitants of the 31 realms, an onward continuation of the processes seeking of Nibbana through death and rebirth of the heavenly deities. The source, or sources, for what is written in the manuscript scripts remains elusive and have not been found. King Lithai's *Traibhumikatha* in Chapter 6 touches on some of the aspects of deity rebirths as does the Ven. Nyanatiloka in his *Buddhist Dictionary*. However, perhaps the inclusion of the selected heavenly texts in the manuscript may rest solely, but unlikely, with the donor of the manuscript Upazin Wizareintasaryi.[LXIX]

The translations which follow have meaning for Theravāda cosmology and in some cases, relevance to the illustrations they are placed with.[LXX] The translations

LXVIII A number of other quotes were vetted, but not included for reason of clarity (see Text 12) and repetition. (See text 2 which while it does amplify Text 1 and present a focus, it also repeats the theme).

LXIX As noted in Part 3, the honorific 'Upazin' indicated that the donor was an ordained monk who, perhaps, made scholarly study and inquiry into other aspects of the cosmos and the continuation of the death and rebirths in the heavens and the states of mind. Perhaps he also thought that the *Traibhumikatha* which presents physical details of the three planes of existence needed amplification. So why not discuss the states of minds of the inhabitants the effects of Jhāna and rebirth in the heavenly realms?

LXX John Guy, in the 1982 December issue of *Orientations* (p. 16) noted, 'An unusual feature of early Buddhist manuscripts was the relationship of illustrations to text; rarely do they illustrate the subject of the text.'

in Parts 3 and 4 of this book, are directly related to what is being shown, mostly the descriptions of the hells, especially the minor ones, have script that describes what will happen if the reader doesn't adhere to established mores, or Buddhists precepts. In contrast each of the heavens illustrated have text placed with them, and as will be seen, many of these commentaries and are not necessarily related to the illustration they accompany. In some cases the text in one place was reinforced by its repetition elsewhere. There also are areas of illegible script, worn away by water, time, and mishandling which cannot be accessed for translation. In the translations, it is evident that there was repetition and duplications. The focus of the text also strayed from the subject and triviality inadvertently occurred as in Texts 3, 13, and others, which give a distance between heavens, when the main subjects are the mind or heavenly rebirths.

Caveats and Conditions

The Pali names of the realms have replaced Burmese ones, to be consistent with other parts of this book. These are shown in **bold** and begin with a capital letter. The Burmese names for the realms as they appear in the manuscript are given in appendix A.

When Burmese nomenclature is used, syllabication of Burmese nouns i.e. Dokgati- ahaik and Thu-gati-haik, favored by the translators, has been kept instead of using Dokgatiahaik and Thugatihaik.

In most cases when Burmese nouns are given it is because Pali, or other nomenclature nouns have not been found. The Pali of Burmese nouns when known are shown in **bold** and in general are in lower case; some exceptions exists, i.e. 'Arahant'.

The word 'person' or 'persons' constantly appears in the following translations. The translations have been confirmed, but the meaning and use of the word is unclear. The heavens are surely blessed with a multitude of inhabitants which negates the use of the words person or persons. However, there may be an explanation of sorts; please see Text 2 and footnote.

It should be noted that the scribes assumed the reader would have general knowledge of the Buddhist cosmos. Terms such as 'the dwellers of the four forbidden realms'; 'ten lower Brahma realms'; and 'the four formless realms' implies that the reader has knowledge of Buddhist cosmological structure.

Unfortunately, the translation of the texts also answers a number of questions that were not asked or beg to be asked. (See Part 3 and the comments on translation.)

The places from where the text has been extracted are shown in the accompanying sketches. See Part 3, 'The Heavens of the Cosmos', for more details.

Table 11 provides an interpretation of the meanings of the Burmese transliterations, not only the celestial minds, but of other possibilities that affect the persona of empyrean worlds. The translations in the table are not complete, nor is the table, and they should be considered provisional and as the beginning of an ongoing inquiry.

TABLE 11. BURMESE TRANSLITERATION AND MEANINGS

Burmese noun, translitera-tion in syllabication format	Provisional Pali equivalent	Meaning of Burmese noun (provisional)	Where found in the texts (T#2=Text 2)
Aeka-besa-Thortarpans.	eka-bījī-sotāpanna	After they became a sotāpanna, those who will be reborn once in any of the 7 sensual realms and will become arahats there. Sotāpanna, i.e. Steam Winner, whose qualities are four: unshakable faith towards the Enlightened One; unshakable faith towards the Doctrine; unshakable faith towards the order; and perfect morality. (Nyanatiloka 1972, p.173)	T#9,
A-haik (clear)	ahetu-[paṭisandhika]	Beings with no root conditions who are born in the four lower worlds or if they are born in the sensuous sphere, are crippled, blind, deaf, mentally deficient, etc.	T#2, T#4, T#10,
A-ku-tho	akusala	No merit, bad and evil deeds.	T#2, T#5, T#10,
Ana-gam	anāgāmi	Non-returners, those who will be reborn in the Pure Abodes of the Brahmā realms and will become arahats there without being reborn in the sensual planes. The honorific for those who discard everything of the physical world and practice, medi-tated according to Buddha's pathway and are non-returners.	T#4, T#8,
Ana-gam mak yar	anāgāmī magga	A type of Anāgāmī mental state; the path stage of a non-returner.	T#8,
Ana-gami-pho	anāgāmī-phala	A stage of enlightenment; the fruition stage of a non-returner.	T#1, T#7,
Ana-gam-mak-pho	anāgāmī-magga-phala	A stage of enlightenment; the path stage of a non-returner.	T#1, T#7,
An-ta-ra- pari- Nirvar-yi-Ana-gam	antarāparinibbāyī-anāgāmī	Honorific of those who have achieved Anāgāmī status in one of the five pure abode realms and who will become an Arahat before they reach the halfway point of their lifespan.	T#8,
Ara-hat-hta-pho	arahatta-phala	Achievement of arahat status; the fruition stage of an arahat.	T#7
Ara-hat-ta-maga-htan	arahatta-magga	A pathway to achieve enlightenment; the path stage of an arahat.	T#1,
A-ri-yar	ariya	A noble, awakened mind.	T#1, T#2, T#3, T#11, T#10,
A-thin-kha-ra-pari-nirvar-yi-ana-gam	Asaṅkhāra-parinibbāyi anāgāmī	A non-returner who reaches nibbāna without exertion.	T#8
A-yu-pa	arūpa	Formlessness.	T#5,
Bong	bhūmi (?)	Realm or residence.	
Deik-hti-ga-ta-weik-pa		The consequence of believing (translation is questionable).	T#7,
Dok-gati-ahaik	duggati ahetu	Realms of suffering where beings with no root conditions are reborn.	T#1, T#4, T#6,
Dwi-haik	dvi-hetu-[paṭisandhika]	Beings born with two positive root conditions (lack of greed and hate).	T#1, T#4, T#6, T#11, T#13,
Ga-na-deik-ta-ya,	[Maybe Ghanika-deva, "Cloud Devas" (?)]	Not known (translation is questionable).	T#8, T#10,

Ku-tho-kri-ya	kusala / kiriya	Tools of making merit; meritorious or inoperative (leading to good results or to no results).	T#10,
Ku-tho-wi-ba-ga-pat	kusala-vipāka	The consequence of meritorious acts.	T#3
Law-kok-tra	lokuttara	Concerning with spiritual world, opposite of lokiya, which mean humanly, or physical world; supermundane, referring to the four paths and fruition stages of noble individuals (ariya-puggala).	T#2, T#10,
Maha-ku-tho	mahā-kusala	Great merit.	T#5, T#8, T#10,
Maha-ku-tho-kri-yar	mahā-kusala-kiriya	The tools of making merit.	T#2, T#8,
Maha-kri-yar	mahā-kiriya	The great tools.	T#8, T#10, T#5,
Ma-naw-dwa-ra-wa-Jhāna	manodvāra jhāna	Skillful mind. [The absorption state at the door of the mind (?)]	T#2, T#5,
Mahek -gok	mahaggata-citta	Exalted mental state of the Fine-Material and Immaterial Jhānas (?)	T#10, T#11, T#13
Nar-nat-hta-ka-ya-nar-nat-hta.	Nava-sattāvāsa	Nine abodes of beings. Nine abodes of beings: (1) Vinipātika ("lower worlds"), beings who are different in body and perception (human beings, some divine beings, some beings in the lower realms; (2) beings different in body but identical in perception (beings born in the Brahmā world through the first jhāna); (3) Ābhassara Devas (Devas of Streaming Radiance) who are identical in body but different in perception; (4) Subhakinha Devas (Devas of Refulgent Glory) who are identical in body and perception; (5) Asañña-satta (Devas who are non-percipient) who are non-percipient and without experience; (6) Ākāsānañcāyatanūpaga Devas (Devas of Unbounded Space) who overcome the perception of forms, etc., and who perceive, "Space is infinite"; (7) Viññāṇanacāyatanūpaga Devas (Devas of the Infinity of Consciousness) who overcome the base of infinity of space and perceive, "Consciousness is infinite"; (8) Ākiñcaññyatanūpaga Devas (Devas of the Realm of Nothingness) who overcome the base of infinite consciousness and perceive, "There is nothing"; (9) Nevasaññānāsaṇṇāyatanūpaga Devas (Devas of the Realm of Neither Perception nor Non-Perception) who overcome the base of perceiving nothing and use the base of neither perception nor non-perception. (Nyanatiloka 1980, pp. 311–12)	T#11, T#13
Oak-dein-thor-tha- Ana-gam	Uddhaṃsota-akaniṭṭha-gāmī	'One who goes upstream to the Highest Realm', Anāgāmīs in the highest of the five realms of Pure Abodes (Suddhāvāsa), who become an Arahat, are called Oak-dein-thor-tha-anagams. The Uddhaṃsota-akaniṭṭha-gāmī (oakdeinthortha Anagamis) can be divided into four types, namely, (1)oakdeinthortha-Akaniṭṭha gami; (2)oakdeinthortha-na- Akaniṭṭha –gami; (3) na-oakdeinthortha-na- Akaniṭṭha -gami and (4) na-oakdeinthortha-Akaniṭṭha -gami. (Pali translation required)	T#8

Pha-la-htan	phalatā	The power or strength of consciousness.	T#13
Pho	phala	Achievement or results.	T#8,
Pok-ko	puggala	A person.	T#7
Tha-thin-kha-ra-nirvar-yi-anagami	Sasaṅkhāra-parinibbāyi anāgāmī	A non-returner who reaches nibbāna with exertion.	T#8
Thaw-thar-pan		First level of consciousness.	T#4,
Thor-tar-pa-ti-mak	Sotāpattimagga	The way or practice of Thor-tar-pan	
Thor-thar-pa-ti-mega-htan	Sotāpanna-magga	A pathway and method to achieve enlightenment.	T#5
Thu-gati-haik	sugati-hetu	A condition for attaining a happy destination.	T#1, T#4,
Thak-ka-dar-gam / Tha-ka-tar-gan	Sakadāgāmi	A Once-Returner who will be reborn only once more in the sensual planes.	T#4, T#9
Thor-tar-pan	Sotāpanna	Stream-winners who have a maximum of seven more lives in the sensual planes.	T#9
Thor-tar-pa-ti-mega- htan	sotāpattimagga	The Stream-winner path.	T#1, T#5,
Thor- tar-pat-ti-mak	sotāpattimagga	The Stream-winner path.	
Thu-gati-ahaik sugati-haik	Sugati-hetu [Same as Thu-gati-haik?]	Good learning mind. [A condition for attaining a happy destination?]	T#1, T#4, T#6,
Ti-haik Pu-htu- Jhāna	ti-hetu putthujjana	Common human being with three root conditions.	T#1, T#2, T#3, T#5, T#10, T#13
Ti-haik	ti-hetu-[paṭisandhika]	Beings born with three root conditions (free from greed, hatred, and delusion) in the human realm or higher realms.	T#1, T#4, T#6, T#11, T#13
U- Ara-hat-hta mak	Arahatta-magga	Questionable; not translated.	T#7,
Viññānaññcāyatana -wi-pat		The requital of Viññānaññcāyatana realm.	T#2, T#5
U-dit-sa	kukkucca	Worry, concern of the mind.	
Upa-hiss-sa-pari-nirvar-yi-Ana-gam	Upahacca-parinibbāyī Anāgāmīs	Those who will become Arahats after they have passed their mid-lifespan are called Upahacca-parinibbāyī Anāgāmīs.	T#8
Wi-pat / Wi-par-ka	vipāka	Resultant, the result of past volitional acts; requital, (you will only get in return according to what you have done).	T#2, T#6, T#7, T#10,
Ya-han-thar	Arahant	The highest level of enlightened individuals, Arahats and Pacceka Buddhas attain final Nibbāna at the end of their lives.	T#1, T#8T#9,

Six Texts Accompanying the Four Heavenly Realms without Material Factors (*Arūpaloka*)

Text 1 deals with the minds of inhabitants of the 31 realms was placed near the illustration of the Nevasaññā-nāsaññāyatana, Realm (1).

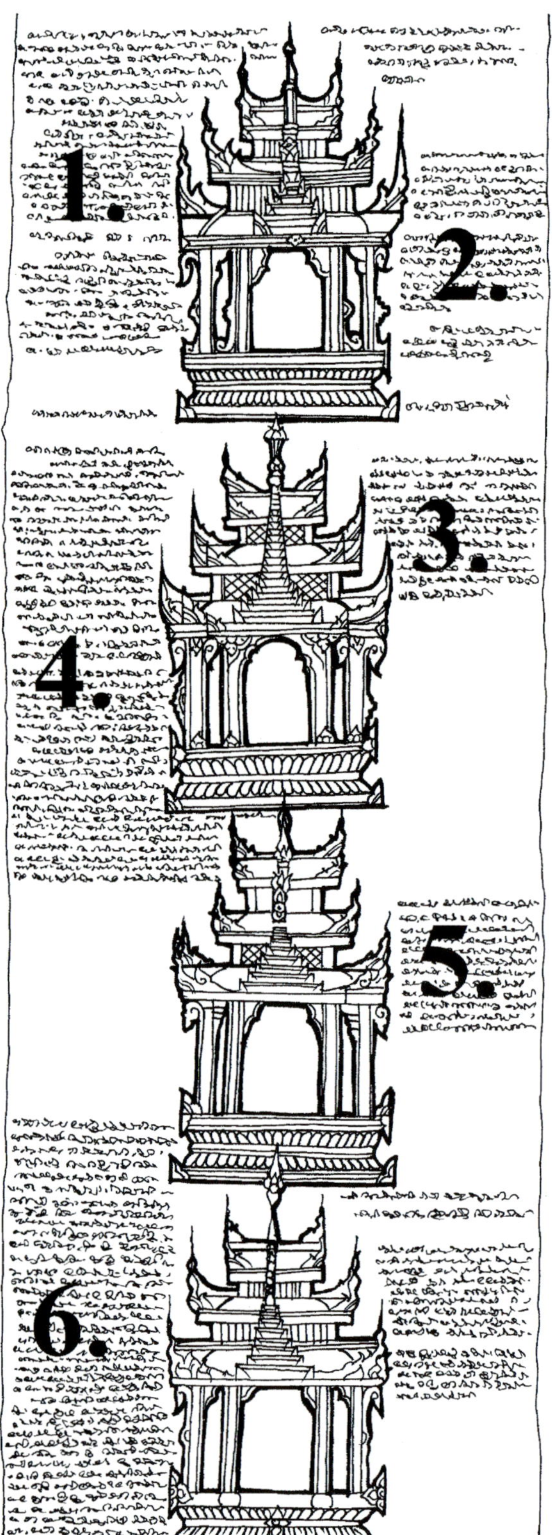

Text Number 1

Of the so called 31 realms of living beings, the dwellers of the four forbidden realms[LXXI] have only **duggati ahetu** minds and in all four realms there is only one person found with this state of mind.

And in the following two realms, namely the human realms and Satumaharit,[LXXII] they have three mind types similar to humans, namely, **sugati-hetu** mind, **dvi-hetu** mind and **ti-heyu** mind and eight **ariya** minds. There are 11 persons in each realm; all together there are 22 persons in both realms.

And of the five realms, namely, **Tāvatimsa, Yāmā, Tusita, Nimmānarati** and **Paranimmita-vasavattī;** they have a **dvi-hetu** mind, a **ti-hetu** mind similar to humans and eight **ariya** minds. All together there are 10 persons in each realm and 50 persons for all five realms.

And of the lower ten Brahma realms.[LXXIII] namely the first level jhāna, second level Jhāna, third level Jhāna, and **Vehapphala**[LXXIV] realms, each has a **ti-hetu** mind and there are eight **ariya** minds, which together constitutes nine persons in each realm and ninety for all ten realms. In **Asaññasatta** they have only a **sugati-hetu** mind with only one person in this realm.

Of the Five Realms of Pure Abodes[LXXV] each have inhabitants who has achieved **anāgāmī-phala**, or who have reached **Arahatta-magga** or who are **Arahant**. There are three for each realm and fifteen for all the five realms.

Of the four formless realms,[LXXVI] there is one being each with a **ti-hetu-putthujjana** mind, and in each realm there are seven beings with **ariya** minds. Therefore, there are eight persons in each realm, or thirty-two persons in all four realms with most suitable minds. Each of the four realms has a being who is blessed with **Sotapanna-magga.**

In all 31 realms there are 218 persons.[LXXVII] That's why it was said, 'Bong-pok-ko-puar-wa-ka-ra'; which means, many persons reside in these realms. (T)

Text 2 deals with the minds of the inhabitants of the Viññānaññcāyatana Realm 3. This text to accomodate space availability is placed near the illustration of the Nevasaññā-nāsaññāyatana Realm (1), the realm of infinite consciousness.

LXXI The realms refer to are Niraya (Level 31), Tiracchāna (30), Preta (29) and Asura (28).

LXXII Manussa (27), the human realm and Cātummahārājika (26) the realm of the kings of the four cardinal directions.

LXXIII The realms referred to are Brahmapārisajja (Level 20), Brahmaparohita (19), Mahābrahmā (18), Parittābhā (17), Appamāṇabhā (16), Ābhassarā (15), Parittasubhā (14), Appamāṇasubhā (13), Subhakiṇṇā (12) and Vehapphala (11).

LXXIV The Vehapphala realm has been identified separately because it is of Jhāna level 4.

LXXV Realms: Avihā (9), Atappā (8), Sudassā (7), Sudassī (6) and Akaniṭṭha (5).

LXXVI Realms: Ākāsānaññcāyatana (4), Viññānaññcāyatana (3), Ākiñcaññāyatana (2), and Nevasaññā-nāsaññāyatana (1).

LXXVII Note: the number of persons do not total 218 persons, but only 211.

Text Number 2[LXXVIII]

This realm is called **Viññānaññcāyatana** because it was clearly defined by the **Viññānaññcāyatana vipaka** mind. The common minds in this realm are: ten **akusala** minds; one **manodvāra-jhana** mind; sixteen **mahā-kusala-kiriyar** minds; twelve formless minds (?); seven clear a-haik minds for **Lokuttara** (concerned with the spiritual world); this is the opposite of Law-ki, which means the human, or physical world, All together there are 41 minds.[LXXIX]

In person there are: one ti-haik-pu-htu Jhāna mind and seven persons with a-ri-yar minds, which together makes eight persons. (T)

Texts 3 and 4 deal with the minds of inhabitants and with the places of rebirth after death. These texts flank the illustration of the Ākiñcaññāyatana Realm (Level 2).

Text Number 3[LXXX]

The uppermost heaven **Nevasaññā-nāsaññāyatana** is 71,860,000 yojana away from the (indecipherable...) human world.

It was named **Nevarsaññā-nāsaññāyatana** because 'it neither is, nor lacks'.[LXXXI]

The beings in **Nevarsaññā-nāsaññāyatana** have thirty-seven minds (concepts?), which are made up of thirty-three minds of **Ākiñcaññāyatana** three **kusala-vipāka** minds and one from **Nevarsaññā-nāsaññāyatana** making a total of thirty-seven. There are eight persons, of which one has a **ti-hetu-putthujjana** mind and seven have **ariya** minds.[LXXXII] (T)

Text Number 4

After the death of living beings with **sugati-hetu** minds, or **dvi-hetu** minds in the human and upper sensual realms, they will be reborn only in the eleven sensual realms[LXXXIII] with the mindset of one of the following four: **duggati ahetu**, **sugati-hetu**, **dvi hetu** and **ti-hetu**, beginning in their embryonic stage. And those who have **ti-hetu**, can be reborn with one of the above four mind sets in these sensual realms, or in realms with other forms or formless high deities, with the exception of rebirth in the Five Realms.

When those with **ti-hetu** minds die from the realms of first stage jhāna, second stage Jhāna, or third stage Jhāna, or the realm of **Vehapphala**, they can be reborn into the formed, or formless high deities, or even in the sensual realms. When they are reborn in the sensual realms, they will have **dvi-hetu** or **ti-hetu** minds but not

LXXVIII This text flanking Nevasaññā-nāsaññāyatana has been placed well above the focused realm.

LXXIX The 41 minds (the minds presented actually add to 46 as per translation) are reserved for beings in this realm, either one or more will be dominant to influence its behavior while others will persist but in an inactive role. Ti-haik pu-htu-zan has the meaning of a human person at the third level of consciousness, but, as there are only minds and no beings in this realm, it may be understood that it could be a deity, of a more human like form but of ti-haik (which is a high rank for a human mind, but of low rank in this realm). Thor-tar-patti is an adjective formed from thor-tar-pan, and thor-tar-patti-mega-htan probably means thor-tar-pan with a highly conscious mind. So it seems that, a person in the human realm could be a person of low birth, high birth or anywhere in between. All are same human beings, but their minds influence the behavior of each person. There is no guarantee that a noble can never be evil and a person of low birth cannot be of great charisma. Each person can only harvest what they have sown in accordance to their deeds.

LXXX The translation is troubled. The phrase 'because it neither has; nor lacks' requires clarification, as does the interplay between Realms 1 and 2. In other sources of the Theravāda Cosmology the four highest realms, those without material factors, have no location-hence no distance from other realms, or the Cakravāla.

LXXXI Perhaps a reference to 'The realm of neither-perception-nor-non-perception'?

LXXXII Asaññasatta Realm (2) and Nevarsaññā-nāsaññāyatana Realm (1).

LXXXIII I.e. Realms 10 to 20.

ahetu minds. Those Thaw-thar-pan, Tha-ka-tar-gan and Ana-gam who died in the realms where deities have form, can be reborn in either the realms of formed or formless deities, from there they can gain enlightenment to enter Nibbana.

Those who die when in the **Vehapphala** realm[LXXXIV] cannot be reborn in higher realms than that.

Those who die when in the *Asaññasatta* realm will not be reborn in the realms of formed or formless deities, but only in the sensual realms alone, because they still have sensual feelings.

From the Five Realms of Pure Abodes – **Avihā**, **Atappā**, **Sudassā**, **Sudassī**, **Akaniṭṭha** – they cannot be reborn in other formed or formless realms, but only return to one of these five realms, where they can find their way to Nibbana.

Those with **ti-hetu** who die in the formless realms[LXXXV] will be reborn with a **ti-hetu** mindset but only in the sensual realms. Those of **sotāpanna**, **sakadāgāmi** and **anagami** from the formless world will return to their own realms and find a way to Nibbana.

Except Athat-nyat-that, the formed or formless realms have a lifespan of 84,000 Mahākappa. (T)

Text Number 5

This text is associated with the illustration of the Viññānaññcāyatana Realm (3) and deals with the minds of beings in that realm.

Viññānaññcāyatana is so called, because it was defined by the **Viññānaññcāyatana** -wi-pat mind. Their minds consists of, ten **akusala** minds that are free of anger; one **mahaggata-citta** mind; eight **mahā-kusala-kiriya** minds; eight **maha-kiriya** minds; twelve **arūpa** minds; seven **sotāpanna-magga** minds making 41 minds in all.[LXXXVI]

In persons there are one **ti-hetu putthujjana** mind and seven **sotāpanna-magga** minds, which is a total of eight persons. (T)

Text Number 6

This text deals with the minds of inhabitants of the 31 realms and amplifies the content of text 1. Text 6 is to the side of the illustration of the Ākāsānaññcāyatana Realm (4).

Of the 31 realms of living beings, the human realm, the realm of six deities, and the realm of the Asura (winipatika asura), when combined are called the 'group of nine realms'.[LXXXVII] This group possesses four types of state of mind since their embryonic stages. These were **duggati ahetu**, **sugati-hetu**, **dvi-hetu** and **ti-hetu**. The inhabitants of these realms have different types of physical features i.e. short, tall, fair, dark, and fat, thin and so on.[LXXXVIII]

LXXXIV Vehapphala is a 4th Jhāna realm.

LXXXV Realms, Ākāsānaññcāyatana (4), Viññānaññcāyatana (3), Ākiñcaññāyatana (2), and Nevasaññā-nāsaññāyatana (1).

LXXXVI The number of minds cited does not tote to 41 as translated, but to 46.

LXXXVII Asura (28), Manussa (27), Cātummahārājika (26), Tāvatimsa (25), Yāmā (24), Tusita (23), Nimmānarati (22), and Paranimmita-vasavattī (21).

LXXXVIII These three types of realms are listed together in a same group and in Pali language it is described as (transliterated) 'Nar-nat-hta-kar-ya-nar-nat-hta'.

The beings of the three first stage jhāna of Brahma[LXXXIX] realms, have several varieties of physical features, but their state of mind since their embryonic stage is only made up of first stage Jhāna **vipaka**. The four forbidden realms only have a single dok-gati-ahaik state of mind in their embryonic stage.[XC]

The three realms of the second stage Jhāna, have two states of mind: the first stage Jhāna **vipaka**, and 3rd stage Jhāna **vipaka**. There are no differences in their features. They are called two **vipaka** group. (T)

Three Texts Accompanying the Heavens of the Five Pure Abodes

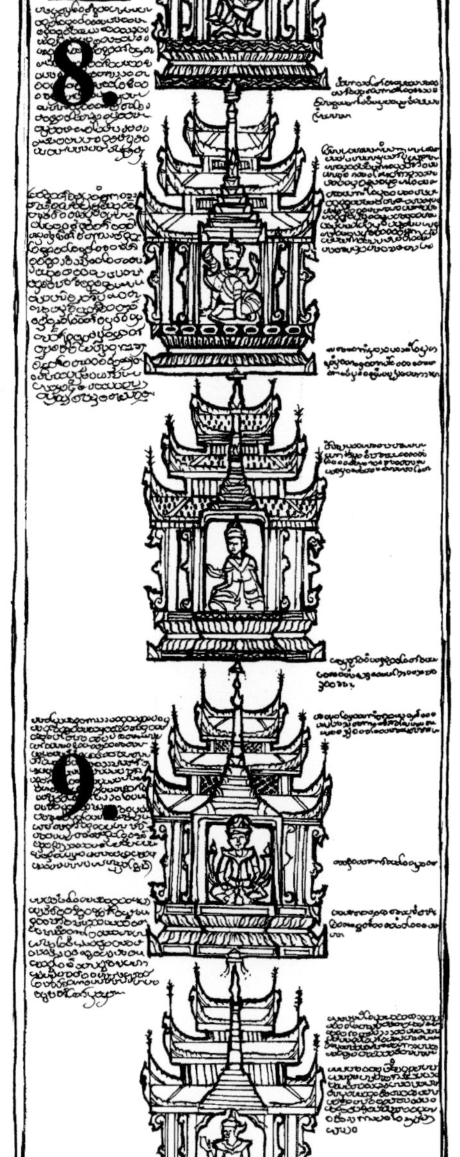

Text Number 7

This text on the right side of the illustration of Akaniṭṭha, Realm 5, addresses the minds of inhabitants in that realm.

It was called **Akaniṭṭha**, because it has the most pleasure to enjoy and is free from any sin. The mind set of the **Akaniṭṭha**, inhabitants consisted of the following mind groups; four deik-hti-ga-ta-weik-pa; one with an **arahatta-maggau**, one with **arahatta-phala**, one **vipaka** of fifth level Jhāna totaling 55. There are three persons [pok-ko] **anāgāmī-magga-phala**, ara-hat-hta-mak and **arahatta-phala**.

The wi-par-ka mind of deceased people who achieved fifth Jhāna is the only one that can be carried to this afterlife by its inhabitants. (T)

Text Number 8

This text on the left side of the manuscript's depiction of Akaniṭṭha, addresses the anagams in the Five Realms of Pure Abodes in great detail. The text requires translation into Pali.

In the Five Realms of Pure Abodes there are five kinds of **anāgāmī**. They are; **Upahacca-parinibbāyī Anāgāmīs**, upa-hiss-sa-pari-nirvar-yi- **anāgāmī**, a-thin-kha-ra-pari-nirvar-yi-ana-gam; tha-thin-kha-ra-nirvar-yi- **anāgāmī**, and **Uddhaṃsota-akaniṭṭha-gāmī**. The **anāgāmīs**, residing in one of the Five Realms of Pure Abodes are those who will become **Arahant**, before they reach the halfway point of their lifespan. There is one with **antarāparinibbāyī-anāgāmī** mind, eight with **mahā-kusala** minds, eight with **mahā-kiriya- anāgāmī**-yar minds, six with ga-na-deik-ta-ya minds, twelve with **ahetu** (clear) minds, eighteen with **maha-kusala-kriyar** minds, one with (indecipherable...)[XCI] pho, one with (indecipherable...). And those who will become **Arahant** after they passed their mid-lifespan are called

Upahacca-parinibbāyī Anāgāmīs. (indecipherable...) They are the **anāgāmīs** in one of the Five Realms of Pure Abodes who are required to spend another life in **Akaniṭṭha** realm before they can realized enlightenment and are called **Uddhaṃsota-akaniṭṭha-gāmī**. The **Uddhaṃsota-akaniṭṭha-gāmī** (oakdeinthortha Ana-gamis) can be divided into four types, namely, (1) oakdeinthortha- Akaniṭṭha gami; (2) oakdeinthortha-na- Akaniṭṭha –gami; (3) na-oakdeinthortha-na- Akaniṭṭha -gami and (4) na-oakdeinthortha- Akaniṭṭha -gami. The **anāgāmīs** who have to spend their lives continuously in every one the four lower realms of the Five Realms of Pure Abodes realms and will become enlightened in the **Akaniṭṭha** realm are called

---

LXXXIX Brahmapārisajja (Level 20), Brahmaparohita (19), Mahābrahmā (18),

XC Therefore these two were listed together in the same group of one. In Pali, it was described as (transliterated) 'Nar-nat-hat-kar-ya-E-kat-hta'.

XCI Believe a word or phrase is missing.

oakdeinthortha Akaniṭṭha- gami ana-gams. Ana-gams. Those ana-gams who have to spend their lives continuously in every one of the three lower realms and become **Arahant** in the Thu-dat-thi (Sudassi) and not reach the **Akaniṭṭha**, are called oak-deinthortha -na- Akaniṭṭha-gami-Ana-gams. The dead **anāgāmīs** from any of the human, lower deities or higher deities realms, who will become **Arahant** in any of the four lower realms of the Five Realms of Pure Abodes are called na- oakdein-thortha-na- Akaniṭṭh*a* -gami-ana-gams. The dead **anāgāmīs** from any of the human, lower and high deities, who will become **Arahant** in the Akaniṭṭha Realm (5) are called na- oakdeinthortha- Akaniṭṭh*a* -gami- **anāgāmīs**. (T)

Text Number 9

This text flanking Akaniṭṭha Realm 5 deals with the rebirth of the thor-tar-pans, those who are on the path to Nibbana. Note: The types of the subdivisions of thor-tar-pans are self-explanatory and have not been included in Table 11.

The **sotāpannas** are divided into two types. They are (1) **sotāpannas** with Jhāna and (2) those who lack Jhāna, or **sotāpannas** without Jhāna.

The **sotāpannas** with Jhāna will become **Arahant** in their same life, whether in the realms of humans or deities. They will never be reborn in the sensual realms again.

The **sotāpannas** without Jhāna are divided into three types: (1) **eka-bījī-sotāpanna** (2) kaw-lan-kaw-la- **Sotāpannas** and, (3) that-tek-khat-tu-pa-ra-ma-thortarpan. After they became **Sotāpanna**, those who have to pass another single life in any of the seven sensual realms and will become **Arahant** are called aeka-besa- **Sotapannas**. Thor-tar-pan, who have to be reborn in any of the seven sensual realms, for one to six times before they can become **Arahant** are called kaw-lan-kaw-la- **Sotapanna**. Thor-tar-pan, who are required to be reborn in any of the seven sensual realms for seven times, are called that-tek-khat-tu-pa-ra-ma-**Sotapanna**.

Sakadāgāmi are also divided into those who have Jhāna and those who lack Jhāna. (indecipherable...) In all, there are five kinds of **sakadāgāmi**. These are;

(1) Those who have become sakadāgāmi in human form and have to pass another rebirth within the human realm to become Arahant.

(2) Those who have become sakadāgāmi in the human realm and have to be reborn in one of the six sensual realms[XCII] to become Arahant.

(3) Those who gain sakadāgāmi in one of the six sensual realms and have to be reborn once in one of these sensual realms before they become Arahant.

(4) Those who gain sakadāgāmi in any of the six sensual realms and have to be reborn once in the human world before they can become Arahant.

(5) Those who gain sakadāgāmi in the human realms, but have to be reborn once in the human realm and then another rebirth in any of the six sensual realms before they can become Arahant.[XCIII] (T)

XCII Cātummahārājika (Level 26), Tāvatimsa (25), Yāmā (24), Tusita (23), Nimmānarati (22) and Paranimmita-vasavattī (21).

XCIII Translation confirmed.

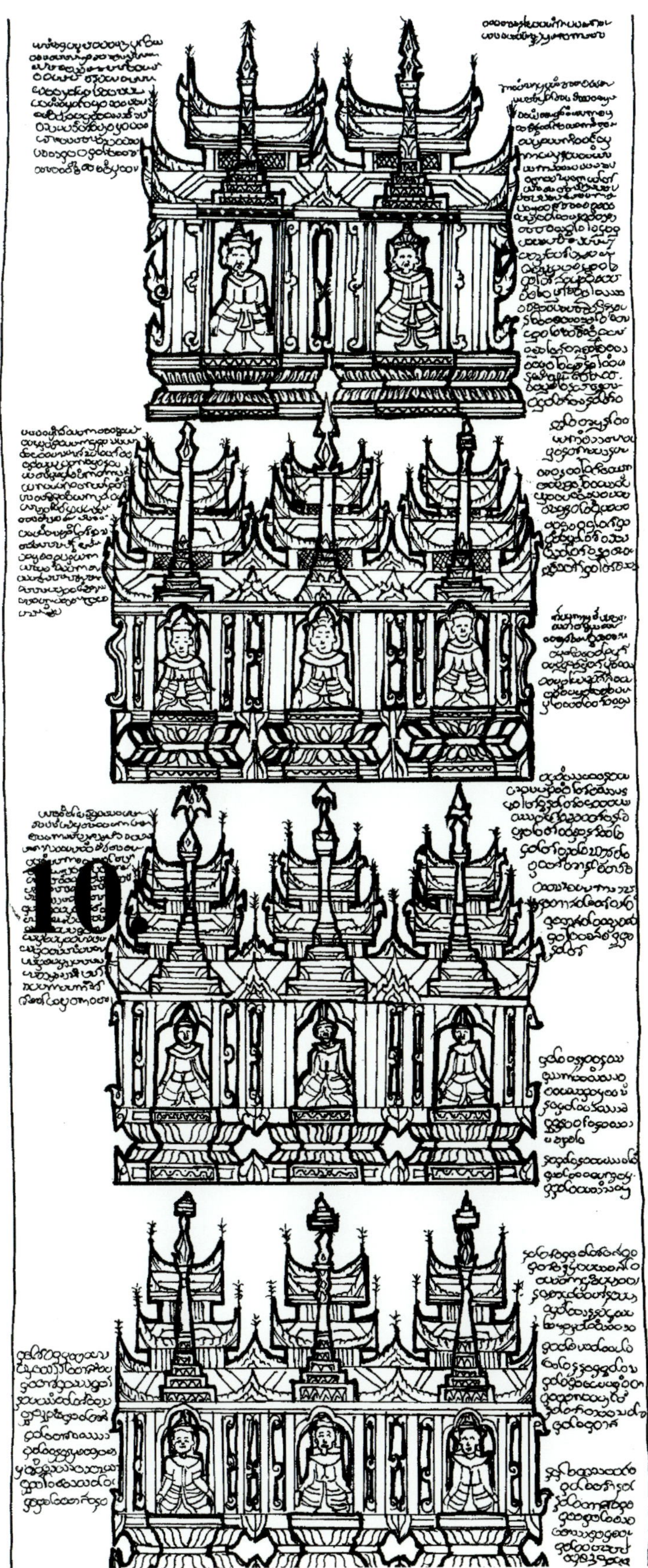

A Text Accompanying Heavens 10 to 20

Text Number 10

This text near the drawing of the *Ābhassarā Realm* (15) deals with the minds of inhabitants of that realm.

Those living in the three realms of second stage Jhāna[XCIV] consisted of the following states of mind: ten clear **akusala** minds; sixteen **mahā-kusala** and **mahā-kiriya** minds; six Ga-na-deik-ta-ya; twelve clear **ahetu** minds; eighteen **mahaggata-citta**, and **kusala-kiriya** minds; eight **lokuttara** minds one second Jhāna **vipāka** mind; and one third Jhāna **vipāka** mind. All together making a total of 66 minds. In the embryonic mind state, they have one second Jhāna **vipāka** and one third Jhāna **vipāka**, thus making an additional two.

In person they have one **ti-hetu putthujjana** mind and eight **ariya** minds, making a total of nine persons which is why in Pali it is stated 'ae-kat-ta-kar-ya-nar-nat-hat.'[XCV] (T)

XCIV These realms are Ābhassarā (Level 15), Appamāṇabhā (16), Parittābhā (17).

XCV Meaning unclear-Pali not found transliteration only provided.

Four Texts Accompanying Realms 21, 22, 23 and 25

Text Number 11

Text 11 on the right side of the Paranimmita-vasavattī Realm (21) deals with the mindsets and lifespans of the inhabitants of that realm.

Paranimmita-vasavattī was named after the pleasure and enjoyment created by others, but belongs with and follows the beings living in this realm. At the end of the six sensual realms the **Paranimmita-vasavattī** realm (21) was reigned over by Mara, the king of evil. Take it in mind to remove nine **mahaggata-citta** minds which should be of concern to the realm and received 80 minds.[XCVI] In person there are one **dvi-hetu** mind, one **ti-hetu** mind, eight **ariya** minds, which makes ten persons in all. Since their embryonic stage, they have [the ability?] and are ready to enjoy splendid pleasure. As they have different varieties of features and appearances, they were called **nava-sattāvāsas**. (T)

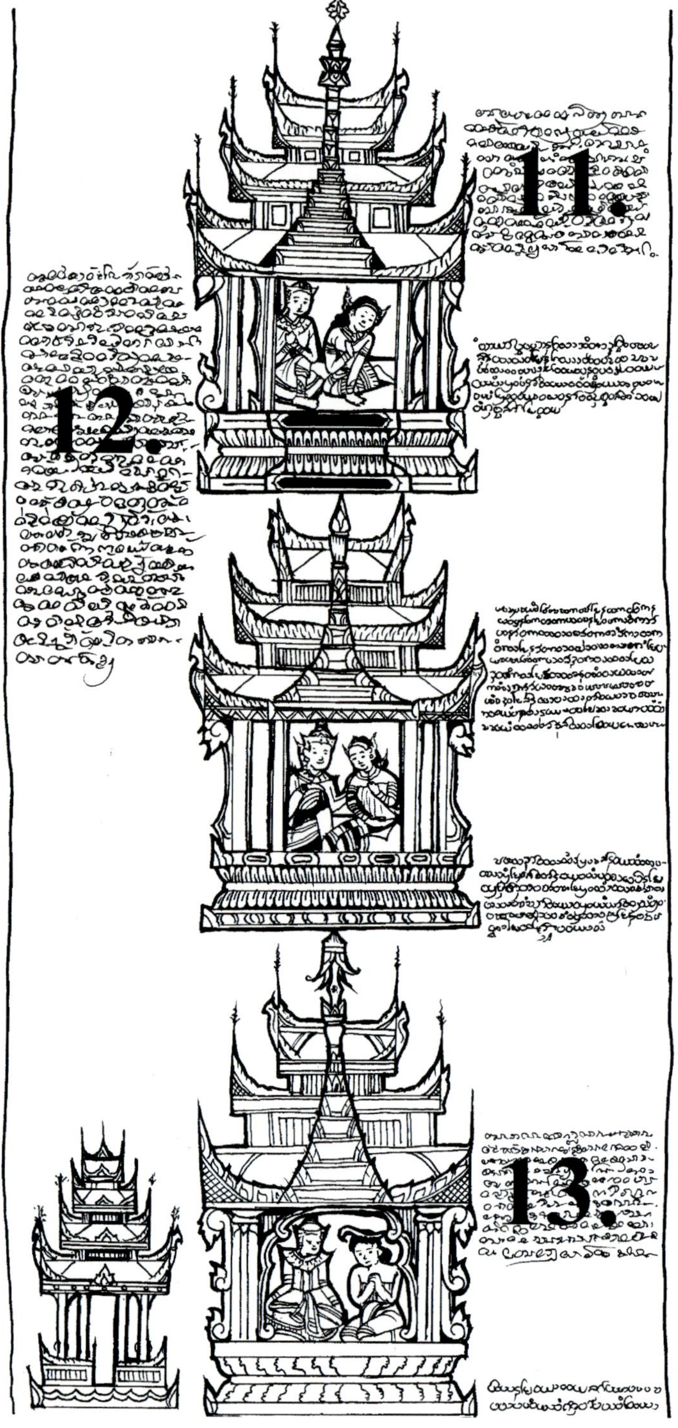

Text Number 12

This text is written in Pali with Burmese script. A partial translation of this rather abstruse text follows.

. . . on a slate write down 1, 2, 4, 8, 16, add 2 zero for the first five and 3 zeros for the rest. This is the lifespan of deities from satumaharit[XCVII] to Para-nein-mita-wa-tha-wa-hti.

Write down 9, 36; 44; (indecipherable...) and 9,216 and add 6 zeros at the end. Compare this figure with the previous figure, calculated with human years and keep in order . . .[XCVIII] (T)

Text Number 13

The texts flanking the right side of the Tusita Realm (23) illustration deals with the minds of its inhabitants.

Because (indecipherable...) it was called Tusita. For beings of this realm, remove the nine **mahaggata-citta** minds and what was left was 80 minds. There are one **dvi-hetu** mind, one **Ti-hetu** mind, four Mega-htan minds, and four **phalatā** minds; in total there are 10 persons in this realm. Since their embryonic stage, they are ready to received all kinds of pleasure, they have many varieties in their feature and appearances. That is why it was called '**Nava-sattāvāsa**'.[XCIX] (T)

XCVI Meaning is not clear but seems to follow a similar pattern to Text 13/.
XCVII Transliterated, meaning not found.
XCVIII The translator noted that text writing became too confusing to continue and abandoned the effort.
XCIX Nine abodes of beings.

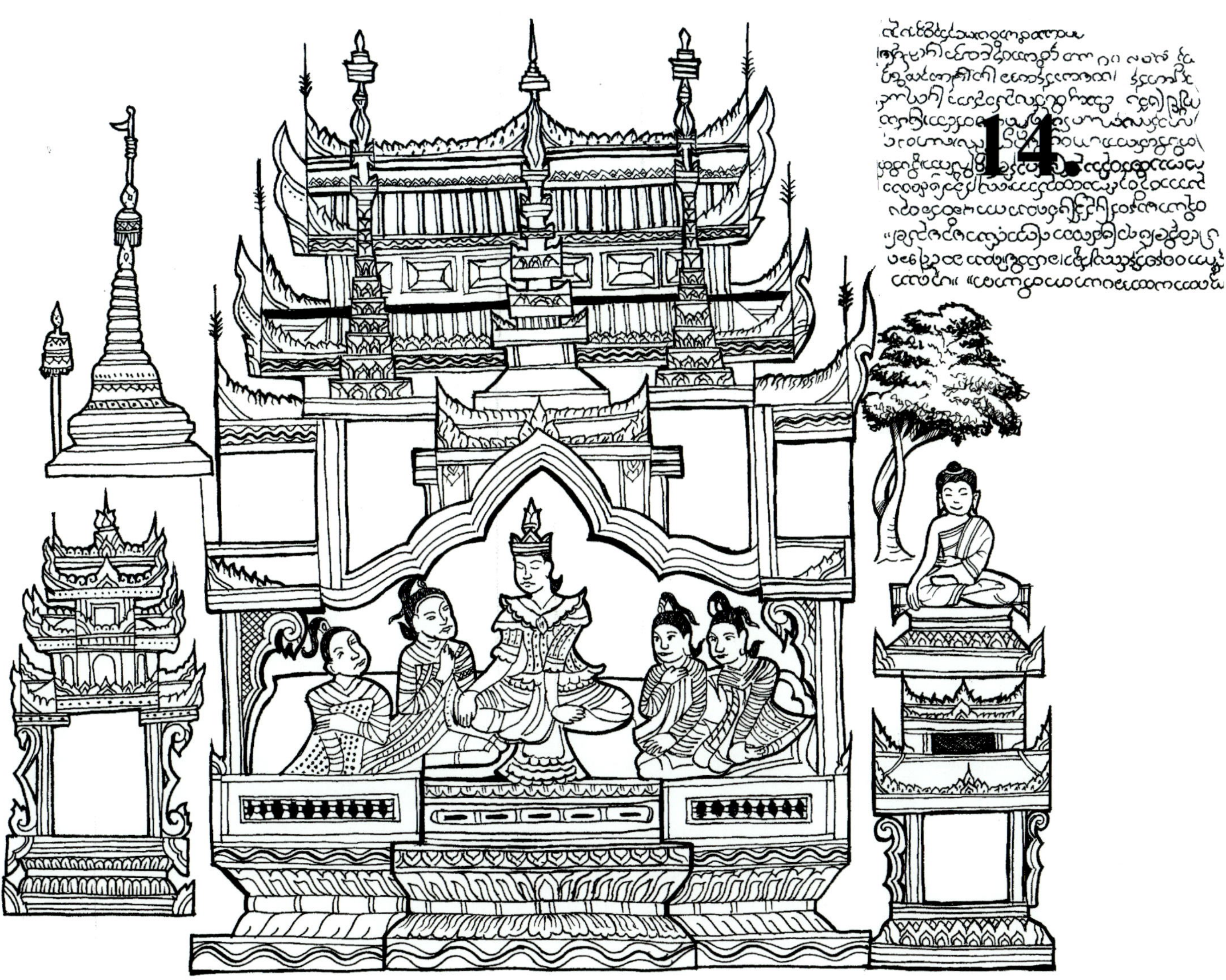

Text Number 14

The text on the right side of Tavatimsa realm (25) illustration deals with the minds of its inhabitants.

It is called Tāvatimsa because it is the abode where the 'Embryo Buddha to be'[C] resides and where the emerald shawl (mya-tha-bet) is to be found. It is the abode of the 32 deities and Indra. Including the 54 sensual minds, they have 80 minds.[CI] From their embryonic stage they enjoy the luxurious pleasure of beautiful splendid abodes, dress and ornaments to enhance their features. (T)

Translator's Notes

Ahaik (**ahetu**) is mostly combined with ***tareikhsan***, with a meaning of 'ignorant animal'; ***haik*** alone was not found in any writing. Since the prefix, 'A-' means negative,

C The translation has been confirmed. There is deviation from other traditions which holds that the Bodhisattva resides in Tustia Heaven. Tusita is the heaven where the Bodhisattva who became the historical Buddha was thought to have resided before being born on earth and where the current Bodhisattva resides. The fact that the Bodhisattva, the future Buddha, 'generally' resides in Tusita heaven is cited in King Lithia's book the Traibhumikatha. The introduction of an additional 54 sensual minds with Indra and his 32 deities does not tote to 80 (though it may be supposed to?).

CI Not understood.

no-haik = ***ignorant***, and ***haik*** could be translated as 'mind', 'concept' or the 'spirit' dominating a person's behavior and activities. Examples of this are**:**

1. ***Dwi haik*** or 'double mind' is a conflicted mind inside a living being, where their behavior will be decided by their mood. In times when good mind prevails he/she will think/ act in good ways, but when the evil mind prevails they will beheve in an opposite manner.
2. ***Ti-hetu****,* or ***Tari-haik***, is the most conscious mind a human being can have; it is the mind reserved for the high deities.
3. All ***haiks*** are gifts to each individual from their birth, and determine whether a person is smart or stupid, good or bad.

Ariya mind can only be achieved by means of hard work, and by practicing total meditation, etc. People who think good and practice in good things are called ***Ariya Pokko***. They follow the 'True eight Makgin pathway'. They stay away from wrong-doings even in their thoughts and words. The eight ***Makgin*** were made of two components, ie. the ***Mak*** (cause/practice/work/the seed sown) and the ***Pho*** (results/gain/harvest).

- ***Kri-yar*** means tools or instruments; in grammatical terms it is a verb.
- ***Ku-tho*** and ***Aku-tho****;* Kutho is good merits, Aku-tho means no good merits, i.e. evil.
- ***Vipaka*** is a requital, i.e. you will harvest only what you have sown. Or, to put it another way your rewards will depend on what you have done.
- ***Sotāpanna*** are those who uphold and practice the true pathway. The doors to the four forbidden realms[CII] are closed for them.
- ***Sotāpanna*** who could discard their greed, anger and neglect may attain a higher level called ***Thak-ka-dar-gam***.
- ***Thak-ka-dar-gam*** (i.e. ***Sakadāgāmi***) who discard everything of the physical world and practice, meditation according to the Buddha's pathway are called ***Anagami***. The highest of all become an enlightened ***Yahanthar*** (i.e. ***Arahant***).
- ***Sotāpanna***, ***anagami***, ***Thak-ka-dar-gam*** are different levels of enlightenment. If we use the term enlightenment for the ***Yahanthar*** (i.e. ***Arahant***), who are ready to enter Nibbana, then the term level of consciousness might be more appropriate instead.

CII These realms are Niraya (Level 31), Tiracchāna (30), Preta (29) and Asura (28).

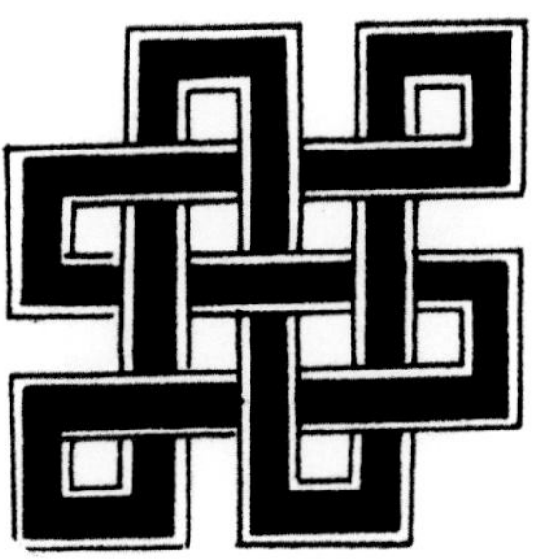

Epilogue

The awesome world of the Theravāda cosmos presented by this Burmese manuscript is difficult to comprehend. In Part 1, which dealt with the Cakravāla, we have been confronted with immensities of space and geological structures that are almost beyond comprehension.

We also have been given dimensions of time which seem to exceed the presence of human life on the earth and perhaps the existence of our known universe by many fold. We are also told that eventually, each world system while subject to growth and development, will, in time, be subject to cycles of decay and periodically destroyed by massive cataclysmic developments of water, fire and wind and then be recreated. These thoughts, concerning the ethical and physical properties of the cosmos, were depicted as reality centuries before King Ruang gathered them together in the sermon of the '*three worlds*'.

Our understanding of Earth's history involves great eons of time from a gathering of dust to its present emergence today – a period that is estimated at 4.5 billion years. Only recently, with the development of the Hubble telescope do we have a comprehension of the development of new galaxies and universes, and the unraveling and destruction of older ones.

It seems that without the advantages of modern science, the people who emerged into the light of knowledge cast by the Buddha almost three millennium in the past foresaw what today's instruments have uncovered Their written thoughts were codified by the work of King Ruang and others. Others have also noticed and commented on the analogous nature of Buddhist cosmology and of modern science.

Winston King wrote:

> Does not science, with its ever-expanding vision of illimitable worlds, many of which may be inhabited, and its ever-lengthening time-scale, accord with the Buddhist cycle theory? Who can longer speak of our little earth as the one and only "world" that has an absolute beginning in time? Is it not far more likely that the history of the physical universe is endless, an infinitely long and beginningless process in which matter or energy forms achieve and lose a patterned existence over and over again; that worlds are born and die without cessation? Such a view, says the contemporary Buddhist, is both Buddhistically and scientifically true at the same time.[225]

The plurality of the worlds of the Buddhist cosmos which are born and then die has been rediscovered by modern science to have reality.

A Polemic for the Theravāda Buddhist Cosmology

The questionable need for a Buddhist cosmology, based on the philosophical character of Buddhist thought, was raised in the Prologue. However, again, Winston King wrote:

> Theravada Buddhism is almost unthinkable without its cosmology, whatever the relation of cosmology and Buddhist practice in original Buddhism. Should one attempt to remove it by some sort of theological surgery, the remaining corpus might well be a corpse instead! And so strong is the rootage of this cosmology in the tradition that the majority of orthodox monks and laity even today would resist any attacks upon or alterations in the Mount Meru cosmology. For them this cosmology represents the clear teaching of the Buddha as interpreted by the orthodox commentators, and this is the end of the matter.[226]

It is agreed: '*and this is the end of the matter.*'

APPENDIX A

REALMS: BURMESE TRANSLITERATIONS TO PALI

Burmese Transliteration	Pali	Realm
Nay-wa-thu-nyar-nar-tha-nyar-ya-ta-na	**Nevasaññā-nāsaññāyatana**	1
Bar-bi-nin-sar-nyar-ya-ta-na	**Ākiñcaññāyatana**	2
Wi-nyar-nin-sar-ya-ta-na	**Viññānaññcāyatana**	3
Ar-kar-thar-nin-sa-ya-na-ta-na	**Ākāsānaññcāyatana**	4
A-ka-ni-ra	**Akaniṭṭha**	5
Thu-dat-thi	**Sudassī**	6
Thu-dat-thar	**Sudassā**	7
Ar-tab-bar	**Atappā**	8
A-wi-ha	**Avihā**	9
A-that-nyat-that	**Asaññasatta**	10
Wei-hap-pa	**Vehapphala**	11
Thu-ba-gi-na	**Subhakiṇṇā**	12
Ap-pa-mar-na-thu-ba	**Appamāṇasubhā**	13
Pa-reik-thar-thu-ba	**Parittasubhā**	14
Bar-ba-tha-ra	**Ābhassarā**	15
Ap-pa-mar-nar-ba	**Appamāṇabhā**	16
Pa-reik-tha-ba	**Parittābhā**	17
Maha-Bya-mar	**Mahābrahmā**	18
Bya-ma-pa-raw-hi-ta	**Brahmaparohita**	19
Bya-ma-pa-ra-thit-za	**Brahmapārisajja**	20
Pa-ri-nein-mi-ta-wa-tha	**Paranimmita-vasavattī**	21
Nein-mar-na-ra-ti	**Nimmānarati**	22
Tu-thi-thar	**Tusita**	23
Yar-mar	**Yāmā**	24
Tha-wa-tein-thar	**Tāvatimsa**	25
Sa-tu-maha-rit	**Cātummahārājika**	26
Ma-nok-tha	**Manussa**	27
Athurakae	**Asura**	28
Byeik-ta	**Preta**	29
Ta-reik-hsan	**Tiracchāna**	30
Nga-ye	**Niraya**	31

APPENDIX B

BURMESE TRANSLITERATIONS FOR COSMOLOGICAL NAMES

Name	Burmese	Description
Anotatta	Ana-wat-ha	The major lake in Jumbudvīpa and source of four rivers
Assakaṇṇa	?	Mountain Range
Avici	Maha A-wi-zi	Hot Hell 8
Cakravāla	Sakyarwalar	The World System
Candra Devaputta	?	Moon
Gandamardana	Gan-da-mar-da-na	Mountain near Anotatta Lake
Godaniya	A-pa-ra-gaw-yan	?
Himavanta	Hi-ma-wun-da	A region of mountains and forest in Jumbudvīpa
Indra / Sakka	Thi-kyar-min	The ruler of Tāvatimsa Heaven
Īsadhara	?	2nd Mountain Range around Mount Meru
Jhānas	Zan, yuzanar	State of Meditation
Jambudvipa	Zam-bu-di-par	South Continent
Kaelarthaba	Kae-lar-tha-ba	Mountain near Anotatta Lake
Kala	Ka-la	Mountain near Anotatta Lake
Kalasutra	Ka-lalthok	Hot Hell 2
Kannamunda	Kan-na-mong-da	1st of the six great lakes of Jumbudvīpa
Karavika	?	3rd Mountain Range around Mount Meru
Kunala	Kong-da-la	2nd of the six great lakes of Jambudivpa
Maha Raurava,	Maha Yaw yuwa	Hot Hell 5
Maha Tapa	Maha-tar-pa-na	Hot Hell 7
Mandakini	Mandak-ini	6th of the six great lakes of Jumbudvīpa
Meemindhara	?	5th Mountain Range around Mount Meru
Mount Meru	Myint-mo	Mountain that is the Axis of the cosmological world
Nimmānarati.	Nein-mar-na-ra-ti	The 22nd level of eistence
Raurava	Yaw-yu-wa	Hot Hell 4
Rhtakara	Rah-tak-ara	3rd of the six great lakes of Jambudivpa
Saddan	sas-dam	5th of the six great lakes of Jambudivpa
Samghata	Thing-gat-tha	Hot Hell 3
Samjiva	The-in-zi	Hot Hell 1
Sattaparibhanda	?	The seven mountain ranges surrounding Mount Meru
Seiktara	Seik-ta-ra	A mountain around Anotatta Lake
Sihappata	Thi-ha-par-ka	4th of the six great lakes of Jambudivpa

Sudassana	?	4th Mountain Range around Mount Meru
Suddhāvāsa	Thok-tha-wa-tha	The Five Realms of Pure Abodes
Suriya-Devaputta	?	Sun
Tapa	Tar-pa-na	Hot Hell 6
Thudatthana	Thu-dat-tha-na	Mountain near Anotatta Lake
Uttarakuru	Oak-ta-ra-gu-ru	North Continent
Vinataka	?	6th Mountain Range around Mount Meru
Yojana	yuzanar/ yu	Measurement of distance in ancient India
Yugandhara	?	1st Mountain Range around Mount Meru

APPENDIX C

THREE VARIATIONS OF THE RANKINGS OF 16–18 ARHATS
Charged with the Continuation and Protection of Buddha's Teachings

A Ranking and Inclusion Listing of the 16 Arhats according to Louis Frédéric[227]

Rank	Name	Manuscript Page Number	Rank	Name (honorific)	Manuscript Page Number
1	Piṇḍola Bhāradvāja	39	9	Jīvaka	
2	Kanaka Bharadvaja		10	Payantakethe	40
3	Kanaka Vatsa		11	Rāhula	41
4	Subuti	42	12	Nagasena	
5	Nālaka	36	13	Angaja	
6	Baddiya	35	14	Vanavasa	
7	Kāludāyi	35	15	Ajita	
8	Vajraputra		16	Chudapanthaka	

A Ranking and Inclusion Listing of the 16 Arhats according to Reginald A. Ray[228]

Rank	Name	Manuscript Page Number	Rank	Name (honorific)	Manuscript Page Number
1	Piṇḍola Bhāradvāja	39	9	Supaka	
2	Kanaka Vatsa		10	Payantakethe	40
3	Kanaka Bharadvaja		11	Rāhula	41
4	Subuti	42	12	Nagasena	
5	Nālaka	36	13	Ingada	
6	Baddiya	35	14	Vanavasa	
7	Kāludāyi	35	15	Ajita	
8	Vajraputra		16	Chudapanthaka	

A Ranking and Inclusion Listing of the 18 Arhats according to Harriet E. Huntington[229]

Rank	Name	Manuscript Page Number	Rank	Name (honorific)	Manuscript Page Number
1	Piṇḍola Bhāradvāja	39	10	Nālaka	36
2	Kanaka Vatsa		11	Kāludāyi	35
3	Kanaka Bharadvaja		12	Vajraputra	
4	Subuti	42	13	Supaka	
5	Nālaka	36	14	Payantakethe	40
6	Baddiya	35	15	Nagasena	
7	Kanaka Vatsa		16	Angida	
8	Kanaka Bharadvaja		17	Vanavasa	
9	Subhinda		18	Ajita	

APPENDIX D

ALPHABETICAL LISTING OF MONK/ARHAT PERSONAGES
with Rank of Importance to Buddhism

Name	Page*	Rank in listing of 16 arhats**	Rank In listing of 10 arhats	Rank in listing of 80 arhats***	Alternative spellings
Ănanda	40		2	30	
Aṅgulimalā	35			76	
Aniruddha	40		8	5	
Assaji	36			45	
Baddiya	35	6		6	Bhada
Baddiya	41			43	
BāKula	40			33	Bakkula
Bawe Tohtay	40				
Dabba	41			25	Malla Putta Dabba
Darusi	40				
Kāludāyi	35	7		32	Kālika
Kimbila	41			46	
Kinkhaā Revata	40		7	15	Khadiravaniya Revata, Katyāyana
Kondadāna	41			22	Kundadhaňa
Kondaňňa	35			1	Aññā Kondaññā
Konhitita	39			29	
Kotikanda Thawana	43				
Kumara	42				
Kumāra Kassapa	36			28	
Lakuddaka Bhaddiya	39			7	
LiPaw The Le	43				
Maha Yakalocaweto	39				
Maha Kāśyapa	38		1	4	Mahākāśyapa
Māhanan	36			44	Mahanama
Mahāsudda	39				
Mandanee Potta Ponna	43				
Mata Buhtay	41				
Mathay	41				An honorific for established monks and not a name
Mawgara Raza	43				

Meghiya	38			67	
Mhakappina	39				
Moggallana	35	---	6	3	
Nadi Kasspa	38			75	
Nāgita	38				
Nālaka	36			49	Nakula?
Nanda	40			37	
Negathen	37				
Payantakethe	40	16		12	Panthaka, Maha. Panthaka
Pilinda Vaccha	36			26	
Piṇḍola Bhāradvāja	39	1		8	Piṇḍolabhāradvāja.
Puṇṇa	41		5	9	
Punnaji	37			74	
Rahttapāla	37			21	Ratthapāla
Rāhula	41	11	10	20	Rāhulā, Rahula
Revata Pandit	42				
Sabhiya	37			79	
Sāgata	38			39	
Saha Udāyi Htera	36				
Sannakha HtayMeinda	40				
Sāriputta	35		3	2	Shāriputra, Sāriputra
Sivali	41			18	
Sobhita	36			34	
Sona	41				
Sona	42				
Subahu Htera	37				
Subuti	42		4	13	Subhuti, Subhinda?
Sulapanna	39				
Take Taha	42				
Taktha Sitay Thij Htay	40				
Tanwara Wana	42				
Taung Phela	42				
Tawka Thuma Htay	41				
Taytahta Htay	42				
Tira	37				
Unknown	42				
Upāli	40		9	35	
Upasena	42				

Upasiwa	41	57	Upasīva, Upasiva
Upavana	38	69	Upavāna
Ūruvela Kassapa	38	31	
Vakkali	35		
Vappa Htera	35		
Vigisa	37	23	Vangīsa
Vimala Htera	37		
Yakalocaweto	39		
Yasa	36	71	
Yasodha	40		
Yawneyhtey	38		
Zartura Pandit	42		

* Refers to manuscript page number

** Source: Reginald A Ray. Buddhist Saints in India. A study in Values & Orientations. Oxford University Press. New York. 1999

*** Eighty Male Arhat Personages Listed by Venerable Piyaratana Māhathera of Ceylon in Rank of Importance to Buddhism

GLOSSARY AND ALTERNATE SPELLINGS

Other spellings include conversion of text usages of simplified Roman alphabet spellings for words from other language sources into more diacritically accurate versions. Also included are variances in the spellings of the same word found in the various texts used as references.

Abhidhamma. (alt. *Abhidharma*). It is used with respect to the Buddha in the jeweled house, which is interpreted to mean the Buddha's beliefs. It is also a term referring to the philosophical and scholastic works of Buddhism found in the Abhidharma-Pitakas of Indian Buddhism.

Abhidharmakośabhāṣyam of Vasubandhu. A Buddhist text written by Vasubandhu an Indian Buddhist and philosopher, circa 319–396 C.E.

Anāgāmin. A person who has overcome the five fetters of life, self, doubt, rituals, and rules sexual desire and resentment, and who will be reborn into one of the heavenly 'pure abodes' and achieve arhatship.

Anguttara Nikaya. The fourth division of the Pali Canon consisting of short discourses arrange with a numerical system to aid memory.

Arhat. (alt. *arahat, arhant, arahant,*) A worthy person; one nearing or obtaining the highest levels of respect and spiritual perfection i.e. a saint who after this life has full extinction and achieves Nibbana.

Attha. (alt. Atta) The soul or self. (*attasarana:* to take refuge in oneself alone.**)**[230]

Bhikshu. (alt. *Bhikkhu)* A mendicant Buddhist monk.

Bodhisattva. (*alt. Bodhisatta*) A person on the road to Enlightenment.

Bodhi Tree. (alt. Bo-tree **)** Ficus Religiosa, a species of fig tree, considered sacred by Buddhist as it is"The tree of Enlightenment"

Brahmā. In Hinduism Brahmā is the first god of the Hindu Trinity of Brahmā, Vishnu and Shiva. Brahmā is thought of as the god of creation and is usually, but not always, shown with four heads and arms.

Brahmin (alt. Brahman) Member of the Hindu priestly caste.

Cakravāla. (alt. *cakkravāla, chakravala)* A 'world system' or an inhabited universe that comes into existence through karmic actions. The cosmos contains a vast number of these nearly identical systems.

Cakravāḍa. (alt. *chakkravāda)* The iron mountain range encircling the *cakravāla* at its outermost edge keeping the great ocean from escaping its bound edge.

Celestial Year. One day in a Celestial Year (C. Y.) in the lower heavens is equal to 50 human years; C. Y. is the equivalent of 18,250 human years.[231]

Chedi. (*alt. Chaitya, Chetya, Chetiya*) An articulated shaped vertical monument, designed around a central axis, used to house relics of holy ones, or ashes of the dead. The shape evolved from burial mounds. Synonymous with the stūpa of Indian Buddhism. Not to be confused with the pagodas of East Asian Buddhism which while also designed around a central axis, have multiple eaves that recede in size from the larger at the bottom of a structure to smaller ones at the top.

Council of Vaiśālī. (*alt. Veśālī*) A council of the Sangha held at Vaiśālī around 100 years after Buddha's death.

Culamani Chedi. The chedi in Tāvatimsa Heaven, visited by Phra Malai, which contains either a hair or tooth relic of the Buddha.

Devata. (*alt. Deva;devatās*) Celestials, heavenly beings, deities who reside in happy worlds above the human plane, but who are also subject to the cycles of rebirths.

Dhamma. (*alt. Dharma*) The nature and condition of things and their essence; religious truths and the doctrines and practices that constitute a religious system. In Theravāda Buddhism the teachings of the Buddha.

Dhyāna. see Jhāna.

Dibbacakku. Supernatural power to see far away things.

Digha Nikaya. Pali text known as '*The Dialogues of Buddha'*.

Dhutanga. List of 13 ascetic practices that prescribe the limits for Theravāda monks pursuing asceticism.

Garuda. Mythical bird composed of human body with the wings beak and legs of a bird. The garuda is the vehicle of Vishnu and the foe of nāgas. The garuda is the royal symbol of Thailand.

Himavanta. a.k.a. Himalayas. Exists in Jumbudvīpa continent with numerous high mountains (80,000–84,000 peaks), great lakes and large forest inhabited by mythological animals.

Indra. Cognomen of Sakka, the lord of Tāvatimsa Heaven and one of the three major Hindu gods.

Isan. Thailand's largest region lying northeast of the capital city of Bangkok; inhabited mainly by ethnic Lao, Khmer and Thai-Khmer peoples. The main economic activity is agriculture.

Jakata (alt. Jātaka) Stories of the previous life of the Buddha. The Jatakas are a collection of 547 stories (Some writers put the number at 550). These stories of the previous lives of the Lord Buddha are thought to be based on pre-Buddhist folk tales. The Jatakas which contained examples of moral approaches to life are a great source of material for teaching the precepts of Buddhism and happily, also a great source of material for the artist. Jatakas are recognized as part of Buddhist literature since they occur in the written canons of Buddhism.

Jhāna. (alt. Dhyāna) Meditative state of profound quiet and utter concentration in which the mind becomes fully absorbed in the chosen object of attention. Jhānas was defined by the Buddha as Right Concentration.

Kalpa. See Mahākappa.

Karma. (alt. *Kamma*) In Theravāda Buddhism, (1) the effects of one's actions and (2) the transmigration of souls from actions and merit accumulated over present and past lives. In Western thought, actions of cause and effect, i.e. 'what goes around comes around'.

Kassapa. One of Buddha's ten great disciples who later, upon the death of Buddha, took over the leadership of the Sangha. He is also known as Mahakāśyapa, i.e. the great Kāśyapa. See following.

Kassapas. The Kassapas of Uruvelā were three ascetics brothers who became holy men. The three Kassapas brothers heard the Buddha espouse the teachings of the Dhamma and they and their followers joined the Sangha.

Khoi Paper. Khoi paper is a product of the fiber of the *streblus asper* shrub, processed by the use of water, pounded into a mash of putty, and then rolled onto a flat surface to produce paper. The result provides a strong durable paper, from which this Buddhist manuscript was produced.

Lithai. King Lithai, the fifth king of the Ruang Dynasty of Sukhothai, the capital of Siam 700 years ago and the author of the *Traibhumikata.*

Mahākappa. (alt. *Kalpa* [Sanskrit] and more common.) Measurement used in ancient India for great and almost immeasurable periods of time. It has been described as the duration of a cosmic period of time of a day and night of the deity Brahmā equal to 4,320 million earthly years.

Mahakāśyapa. See Kassapa.

Mahāyāna. One of the two great schools of Buddhism; identified as 'the greater vehicle' This branch of Buddhism, which espouses reliance on the bodhisattvas for the seekers of Nibbana began in India at the beginning of the common era and spread mainly throughout East Asia.

Maitreya. (alt. Metteya) The future Buddha.

Māra. (alt. Mara)An evil deity; lord of the 6th heaven of sensuous desires and wants. Māra symbolizes the passions which overwhelm humans and stops good from coming forth. Māra attacked the Lord Buddha in his quest for enlightenment for a period of six years before the Buddha achieved his transcendence.

Moggallana. (alt. Moggallāna, Maudgalyāyana, Mahāmaudgalyāna, Mahāmoggallāna). One of two chief disciples of Buddha. In paintings of the Buddha and his two disciples Moggalana is always depicted on the Buddha's left side. Acknowledged by the Buddha to have attained great miraculous powers.

Mount Meru. (alt. Sumeru, Simeru). The Buddhist world cosmological system is based on the great axial mountain of Meru. The top of the mountain is the site of the Tāvatimsa Heaven, Indra's celestial city.

Nāga. Mythological cobra-type serpent, sometimes with multi-heads in which form it served as protector of Buddha During his meditations after Enlightment (muchinda). The nāga's foe is the garuda.

Naraka. See Niraya.

Nat. A Burmese spirit or god. The nats are a mixture of Hindu gods, prominent personalities who have died, and others. There are 37 recognized nats.

Nibbana. (alts. Nibbāna, Nirvana, Nirvāṇa) The goal of Buddhism. The extinction of rebirth and future existence.

Niraya. The nether or infernal world; 'Hell'.

Noble Order. Consist of the 80 Male Arhats and 13 Female Arhats who by their efforts belong to the inner circle of Gautama Buddha.

Pacceka Buddha. (alt. Pratyeka Buddha) A solitary religious person who achieves enlightenment by his own doing

Parabaik. In Burma, a 19thcentury handmade book consisting of leaves of Khoi paper, folded as an accordion.

Phyathat. In Burma, the tiered roof and clearstory of a religious building.

Phra Bot. Hanging cloth painting of any size, normally for occasional use in temples.

Preta. (alt. Peta) 'Hungry Ghosts'A miserable existence; Realm 29 of the 31 realms of the cosmos.

Rattankosin. The name given to the new capital, Bangkok, by Rama I in 1782; also identifies a school of Thai art.

Rishi. Holy person, ascetic, seer, saint, poet, or hermit. Refers also to the seven rishis to whom the Vedic hymns were revealed.

Ruang. The name of the 14th century dynasty in the kingdom of Sukhothai, Thailand

Sakka. Cognomen of Indra.

Sala. Thai or Lao, open walled, pavilion.

Saṃsāra. Cycles of endless rounds of death and rebirth in which humans and other sentient beings make their existence. Humans can affect their karma and thus migrate from one status to another, either better or worse, in the cycle of rebirth and death – depending on the cumulative results of their day-by-day volitional actions.

Sangha. Community of Buddhist monks.

Sariputta. (alt. Śāriputra, Sāriputta, Shāriputra) One of two chief disciples of the Buddha. In paintings of Buddha and his two disciples Sariputta is to the Buddha's right. He is acknowledged by the Buddha as the foremost of all disciples in the attainment of wisdom.

Sattaparibhanda. The term for the seven mountain ranges surrounding Mount Meru

Siddhartha. (alt. Siddhattha, Siddhārtha) Given name of the historical Buddha. Siddhartha Gotama. Siddhartha means one who has accomplished set goals.

Stupa. See Chedi.

Sūtras. (alt. Sutta)Sermons or discourses attributed to Buddha or a disciple.

Tāvatimsa Heaven. (alt. *Trāyastriṃśa,, Tāvatimsa, Trayastrimsha*) The 25th level heaven; 'he abode of the god Indra, and the 32 gods ruled over by him in

Tempera. A method of painting which uses an albuminous vehicle, such as egg yoke, instead of oil, as a binder.

Theravāda. 'Doctrine of the elders', the oldest form of Buddha's teachings handed down using the Pali language. The Buddhist philosophy as practiced in Sri Lanka and Southeast Asia.

Thin. In Burma a bird's head finial positioned at the end of the peaks of Thai religious structures. By tradition said to be either, a Garuda, the vehicle of Vishnu, or a Hamsa (a goose or swan), the vehicle of Brahmā.

Tosachāt. The last ten of the Jatakas. Buddhist birth stories which are illustrative of the virtues required for the future Buddha.

Traiphum. (alt. *Trai Phum, Tryphuum*) The three-tiered hierarchical worlds' structures of Theravāda cosmology.

Triloka. See Traiphum

Usuda. Minor or auxiliary hell

Vaijian. Peaceful people whose enlightened social and political practices were suggested by the Buddha as a model for the Sangha. After Buddha's death ten lax practices by monks lead to a convening of the First Buddhist Council.

Vedas. A large compilation of the complex and oldest texts of Indian literature which Hindus believe to be documentation for divine authority and superhuman origins. The oldest of these texts is believed to have been written around 1500–1200 B.C.E.

Vinaya Pitaka. One of three sections of the Theravāda Pali Canon, dated to 1st century BCE that lists 227 rules applicable to monks. An additional section for nuns is included.

Wihan. An assembly hall in a temple compound.

Wat. Buddhist temple and compound in Thailand, Laos, and wherever Tai-speaking people practice Theravada Buddhism. The wat compound also includes ancillary religious and domestic structures and uses.

Yaksha. (alt. Yakkha) Ogre both good and bad; guardians of temples. Yakshas are often portrayed and seen as giant, multi-colored statues at entrance doors of Thai temples

Yāmā. Yama, ruler of the hells. He is in charge of the karmic process; as dead souls are reviewed he reminds them that they alone are responsible for being in hell and tells them of the punishment for their actions and offences.

Yojana. Measure of distance in ancient India between 4 and 13 kilometers.[232]

EXPLANATIONS OF SYMBOLS

Lotus: One of the primary Buddhist symbols, the open lotus flower resting on a pond's surface, with roots in the muck, represents the true nature of beings untouched by the mud of the world; purity of mind. Immediately following his birth the Buddha took seven steps; wherever his feet touched the ground a lotus bloomed.

Infinite Knot: In Buddhism the Infinite Knot is an indication that everything in this life is interconnected, reflecting the endless cycle of death and rebirth.

Chakra: The wheel of the law. This is the symbol of the doctrine that Buddha enunciated with the first sermon. The Chakra is the symbol of the eternal cycle of birth, death, and rebirth, as well as of the constantly Turning Wheel that the Buddha set in motion when he gave his first sermon.

Triratna: The three jewels of Buddhism:

The Buddha

The Dhamma-(the teachings)

The Sangha –(the community of monks)

Pharasa: The battle axe which symbolized the severance or cutting away of wealth and worldly attachments

ENDNOTES

Prologue

1 Reynolds & Reynolds 1982, p. 35.

2 Sadakata 2009, p. 181.

3 See: Blanchard 1958, p. 99.

4 Bechert & Gombrich 1984, p. 9.

5 Berry 1975, p. 57.

6 Lopez. 2001, p. 219.

7 Berry 1975, p. 57.

8 See: Frédéric 1995, p. 19.

9 Reischauer & Fairbank 1960, p. 144.

10 Lopez 2001, p. 263.

11 de Bary 1972, p. 9.

12 Lopez 2001, p. 263.

13 See: Tilakaratne 2012.

14 Tiyavanich 2003, p. 318.

15 Lithai 1985, p. 11.

16 Reynolds & Reynolds 1982, p. 5.

17 B.C.E. = Before Common Era (equivalent to B.C.); C.E. = Common Era (equivalent to A.D.).

Part 1. Buddhist Cosmology

18 Reynolds & Reynolds 1982, p. 15.

19 Knappert 1995, p. 10.

20 Ions 1967, p. 13.

21 Ions 1967, p. 13.

22 Ions 1967, p. 14.

23 Ions 1967, p. 15.

24 Ions 1967, p. 23.

25 Wilkins 1980, pp. 93–98.

26 Ions 1967, p. 41.

27 Ions 1967, p. 88.

28 See Ions 1967.

29 Nyanatiloka 1972.

30 Ions 1967, p. 135.

31 See: Snelling 1998, pp. 26–28.

32 Reynolds & Reynolds 1982, p. 15.

33 Jhāna is discussed in greater detail later in Part 1.

34 See Armstrong 2001, p. 94.

35 Author's collection. Pen and ink drawing extracted from a pigment and gilt painting on cardboard, Thailand Rattanakosin Style. circa 1850–1870. 32.5 cm x 42 cm. Quote is from USIS, 1957. Picture 47.

36 Author's Collection. Figures and drawing extracted from a Phra Bot. circa 1990.

37 Sources: Jain cosmology-Wikipedia; Caillat & Kumar 1981.

38 Wilkins 1980, p. 488.

39 Author's Collection. A Jain Painting Of Jumbudvīpa**.** Opaque Pigments and Ink on Paper. Rajasthan School. Circa 19th Century (1810–1825). 24.1 cm. x 26 cm / 9.5" x 10 ¼"

40 Sources: Christie's Catalog, 20 March 2012, New York. *The Doris Wiener Collection*. Item 329; Caillat & Kumar 1981.

41 Snelling 1998, p. 43.

42 Sadakata 2009, p. 11.

43 Kloetzli 1983, cover; inside flap.

44 See Lopez 2001, p. 19.

45 See Lopez 2001, p. 22.

46 Reynolds & Reynolds. 1982, p. 27.

47 Reynolds & Reynolds 1982, p. 16. The Reynolds & Reynolds book introduces 'World and Realm' as a substitute for *bhumi.*

48 Vasubandhu 1991.

49 Reynolds & Reynolds 1982 and Lithai 1985.

50 Powers 2000.

51 "*said to be the distance that one can travel on horseback between harnessing and unharnessing a horse. There are various estimates of how far this is ranging from four to thirteen kilometers*" (Powers 2000).// "*It is based on the distance that an army can march in a day, i.e.15–20 kilometers"* (Schuhmacher, Stephan & Woerner, Gert, eds. 1999). // "*There are various explanations concerning the length of an yojana; one says it is about seven kilometers."* (Sadakata, 2009, p. 25). //One author eliminates the conversion factor of yojanas directly replacing one yojana with one mile, or 1.6 km/yojana (Evans-Wentz, 2008, p. 48). There are also Burmese and Thai measurements which use the yojana today. These are contemporary, and while valid, and the name may be derive from ancient writings they bear no direct lineage to the measure of distance used in ancient India.

52 The relationship of wa/yojana has been deduced from figures given by Reynolds & Reynolds 1982. On pp. 238 and 239. The figure 2,688,000,000 *wa* is given here and it is stated to be 336,000 yojana which is by deduction 8,000 wa per yojana.

53 *Geographica* 2004, p. 11.

54 Sources: Keown 2004. Also see Schuhmacher, Stephan & Woerner, Gert 1999.

55 Definition for Celestial Years in most Buddhist reference books (see references) is not to be found.

56 Keown 2004.

57 Wikipedia. http://en. wikipedia.org/wiki/ *Dhy %C$%81na_in_ Buddhism* (accessed 6/24/2013).

58 Reynolds & Reynolds. 1982, p. 359.

59 Wikipedia. http://en. wikipedia.org/wiki/ *Dhy %C$%81na_in_ Buddhism* (accessed 6/24/2013).

60 Wikipedia. http://en. wikipedia.org/wiki/ *Dhy %C$%81na_ in_Buddhism* (accessed 6/24/2013).

61 Reynolds & Reynolds 1982, p. 57.

62 The table has been developed from (1) direct translations and (2) a fusion of sources:

Herbert 2002, pp. 77–98.

Vasubandhu 1991.

Lithai 1985.

Reynolds & Reynolds 1982.

Wikipedia. http://en.org/wiki/*Buddhist cosmology* (accessed 12/31/2010).

63 Herbert 2002, p. 92.

64 Reynolds & Reynolds 1982, p. 5.

65 The distances from the heavens to the ocean vary greatly with different sources. Huntington's & Bangdel's book '*Circle of Bliss*' in the section which dealt with the 'Mount Meru System' contained the most expansive numbers. Figures given in Wikipedia (http://en.wikipedia.org/wiki/ *Buddhist_ cosmology* [accessed 3/11/2012] were reduced a thousand fold.

Huntington's & Bangdel's figure for level 5 (*Akaniṭṭha*) distance above the ocean was 335,544,320,000 yojana.

Wikipedia's figure for level 5 (*Akaniṭṭha*) distance above the ocean was 167,772,160 (a figure that bears resemblance to Huntington's & Bangdel's figure for level 6 *Sudassī of 167,772,160,000*).

Figures given in Reynolds & Reynolds, (p. 241) for four heavens only approximated those of Huntington's & Bangdel's figures.

Aaegi	Reynolds' & Reynolds'	Wikipedia's	Huntington's & Bangdel's
Level 21	1,344,000 yojana	1,280 yojana	1,280,000 yojana
Level 22	672,000	640	640,000
Level 23	336,000	320	320,000
Level 24	168,000*	160	160,000

*includes Mount Meru's 84,000 yojana height

It seemed appropriate given the awesome dimensions of the totality of the cosmic universe to use Huntington's & Bangdel's figures.

66 The distances from the heavens to the ocean vary greatly with different sources. It seem appropriate given the awesome dimensions of the cosmic universe to use the largest.

67 Wikipedia, http://en.wikipedia.org/wiki/Preta; 2/27/2012.

68 Author's collection. Rattanakosin style. Circa: 1st quarter 20th century. Pigments on Khoi paper. 29 cm x 18cm./11.5" x 7.0".

69 Author's collection. Rattanakosin style. Circa: 1st quarter 20th century. Pigments on Khoi paper. 29 cm x 18cm./11.5" x 7.0".

70 Author's collection. Rattanakosin style. Circa: 3rd quarter 19th century. Pigments on Khoi paper. 29 cm x 18cm./11.5" x 7.0".

71 Snelling 1998, p. 47.

72 Lithai 1985, p. 47.

73 Vasubandhu 1991, p. 459.

74 Alabaster 1871, p. 14.

75 Sadakata 2009, p. 12.

76 Sources: The diameter of the Cakravāla is at best elusive. Sadakata gives a figure of 1,203,450 yojana (Sadakata 2009, p. 27.

Kloetzli (1983, p. 26) gives a figure of 1,120,000 yojana. Since Sadakata uses 8 km/yojana and this book uses 4 km/ yojana the only relativity to the question of size of the diameter of the Cakravāla is an order of magnitude. A diameter of 1,200,000 yojana has been selected for reasons of clarity of presentation. The earth's diameter is from *Geographica* 2004.

77 Source for Cakravāḍa: Sadakata 2009, p. 26.

78 Sadakat 2009, p. 29. The figure has reduced slightly from 322,000 yojana to 320,000 yojana for clarity of presentation.

79 Keown 2004, p. 48

80 Grogerley 1908.

81 Sadakata 2009, p. 29.

82 Jusmai 1989, p. 127.

83 Rooney 2005, p. 130.

84 Reynolds & Reynolds 1982, pp. 275, 276.

85 Vasubandhu 1991, p. 463. The size is hedged at either 20,000 or 80,000 yojanas on a side. However if it were to be 80,000. yojanas that would be the same as the base and would not allow for not inclinations on its way to the summit and, hence, no terraces. What is left is a very large, straight-sided column.

86 Vasubandhu 1991, p. 462.

87 Huntington & Bangdel 2003, p. 67.

88 Reynolds & Reynolds 1982, p. 275.

89 Reynolds & Reynolds 1982, p. 275.

90 Reynolds & Reynolds 1982, p. 275.

91 Reynolds & Reynolds 1982, pp. 223, 224.

92 Sadakata 2009, p. 29.

93 Reynolds & Reynolds, p. 276, 277.

94 Sadakata 2009.

95 Lopez 2001, p. 21.

96 The earth's diameter is 7,925 miles or 12,756 kilometers or 3,185 yojanas

97 Source of figure (modified): Grogerley 1908.

99 Vasubandhu 1991, p. 455.

100 "This *Jambu* continent of 10,000 yojana from east to west is of a shape similar to the (triangular) shaft of an ox cart" (Lithai 1985, p. 411). Reynolds & Reynolds reflected the foregoing and noted, "This sentence is problematical since it seems to imply that Jambu continent and the faces of its inhabitants have a triangular shape rather than the round shape that is clearly affirmed in two previous references (see Chapter 5 and n. 35 above)" (Reynolds & Reynolds 1982, p. 303, fn. 67).

101 Sadakata 2009, pp. 32–38.

102 Source: Lithai, King. op.cit, pp. 395–411.

Part 2. The Manuscript: An Overview

103 See McDaniel 2009 for a discussion of this practice.

104 Herbert 1999, p. 89.

105 Herbert 1999, p. 92.

106 Source: Herbert 1999, p. 90.

107 Herbert 1999, p. 92.

108 Herbert 1999, p. 92.

109 Herbert 1999, p. 93.

110 Herbert 1999, p. 93.

111 Garnier 2004, p. 139.
Source: Herbert 1999.

112 Lithai 1985.

113 Bogle 2011, p. 138.

114 Herbert 1999, p. 93.

115 *Trai Phum Book* 1999, Volume 1, Picture 39.

116 Herbert 2002, p. 80.

Part 3. Manuscript: The Worlds of the Cosmos

117 While King Lithai's Traibhumikatha order of presentation begins with the hells and moves towards the heavens this is not the case with Thai and Burmese cosmological manuscripts which begin with the heavens. Trai Phum Book: Ayutthaya Manuscripts–Thonburi Manuscripts. 3 Volumes. Amrin Printing & Publishing Company. Bangkok. 1999 (Thai language) Book published in honor of the King's 6th cycle (72nd year).

119 Source: Translation.

120 Herbert 2002, p. 80.

121 Source: Bodhi 1999.

122 Keown 2004, pp. 242, 243.

123 Herbert 2002, p. 83.

124 Herbert 2002, p. 81.

125 Herbert 2002, p. 82.

126 Fischer-Schreiber et al 1999, p. 388.

127 Herbert 2002, p. 82.

128 Source of drawing: author's collection. Original painting, pigments on Khoi Paper. Thailand. Rattanakosin School Circa Mid 19th Century. 37 cm x 28 cm /14.5" x 11.0".

129 Herbert 2002, p. 82.

130 The reproduction is the work of Theeravut Sernkrang. Basement Night Bazaar,Changklan Road, Chiang Mai.(Circa 2011)

131 Shaped like a stylized bird's head it symbolizes either the Garuda, the bird vehicle of Vishnu or the Hamsa, the vehicle of Brahma. Source: Chaturachinda Krishnamurty & Tabiang 2004, p. 59. As transliterated from Burmese 'thin'

132 Herbert 2002, pp. 82–83.

133 Author's collection. Pigments painted **on** Khoi Paper. Thailand. Rattanakosin School Circa Mid 19th Century. 37 cm x 28 cm /14.5" x 11.0"

134 Brereton 1995, p. 193.

135 Source of names and descriptions; Reynolds & Reynolds 1982, p. 219.

136 Herbert 2002, p. 93.

137 Sources: Wikipedia, http:en.wikipedia.org/wiki/Preta 2/27/2012; Reynolds & Reynolds 1982. Chapter 3.

138 Author's Collection. Primitive Style, Northeast Thailand / Cambodia. Circa: 3rd quarter 19th century. Pigments on Khoi paper. 29 cm x 18cm./11.5" x 7.0" Note the Northeast area of Thailand is also known as 'Isan'.

139 Author's collection. Rattanakosin style. Circa: 1st quarter 20th century. Pigments on Khoi paper. 29 cm x 18cm./11.5" x 7.0". (The Phra Malai story is one of the core texts in Theravāda Buddhism in Southeast Asia. The story may have originated in Sir Lanka and 'gained legs' as it was told and retold over the years, moving through the countries of Southeast Asia. The story of Phra Malai flowered in Thailand and Cambodia in the 18th and 19th centuries. Phra Malai, a saintly monk, an *arhat,* through meditation gained the powers to move between heaven and hell and to travel through various cosmic realms. In some Theravāda accounts he was Moggallana's successor).

140 Snelling 1998, p. 45.

141 Keown 2004, pp. 222, 223.

142 Sources. Keown. Op.cit. Schuhmacher, Stephan et all.1999.

143 Reynold & Reynold 1982, p. 68.

144 Vasubandhu 1991, p. 456.

145 Lithai 1985, 71.

146 Ibid.

147 Reynolds & Reynolds 1982, pp. 80–81.

148 Reynolds & Reynolds 1982, p. 81.

149 Reynolds & Reynolds 1982, pp. 81,82.

150 Trai Phum Book 1999. Volume 1, Picture 37.

Part 4. Cosmography

151 Alabaster 1871, p. 16.

152 Herbert 2002, p. 85.

153 Herbert 2002, fn. 43, p. 96.

154 Herbert 2002. Figure 8.6 (p. 87) shows a similar located figure of a nāga.

155 Reynolds & Reynolds 1982.

156 Sadakata 2009, p. 31.

157 Vasubandhu 1991, p. 455. There are a number of approaches to the development of the geography of Jumbudvīpa; one, a Jain version, was presented in Part 1. Reynolds & Reynolds presents a line diagram without scale (page 362) and with north to the top of the page. In this diagram the Ganga River flows south. The orientation of Jumbudvīpa in Professor Akira Sadakata's book is not clear. Professor Sadakata also uses the configuration from the *Abhidharmakośabhāṣyam* and has north to the bottom of the page with the Ganga River flowing west (page 33).

158 Ibid.

159 Herbert 2002, p. 87.

160 Reynolds & Reynolds 1982, pp. 295–296.

161 Source: Lithai 1985, pp. 397, 399, 401.

164 Ibid.

165 Herbert. op.cit, p. 87.

166 Source: Nyanatiloka 1972, p. 119.

167 Trai Phum Book: Ayutthaya Manuscripts–Thonburi Manuscripts 1999. Volume 1, p. 73.

168 Translation by Ashin Sopaka (see acknowledgements).

169 This illustration in the manuscript has the north towards the top of the paper varies with that of Figure 33 which has the north at the right side of the paper; both hew to current convention of placing north in illustrations either to the top or to the left of the paper.

171 Reynolds & Reynolds 1982, p. 291; Lithai, King 1985, p. 395.

172 Source:Burmese scholar and translator.

173 Reynolds & Reynolds 1982, p. 291; Lithai, King 1985, p. 395.

174 Source:Burmese scholar and translator.

175 Source: www.sacredsiam.com/thailand-amulets/nareepol-tree-fruit-women.html. 2/24/2014.

176 Trai Phum 1999. Volume 1, page 185.

Part 5. The Arahats

177 See: Pal 1990, p. 68.

178 Snelling 1998, p. 81.

179 King 1964, pp. 55 & 70. King notes what others also have expressed "... instruction, which might well include missionary teaching, is part of the monkish discipline or duty, the monk's concern to achieve his own Nibbana tends to neutralize such effort"(55) and "For the monk upon entry in to the *Sangha* dedicates himself to the direct pursuit of Nibbana"(p. 70).

180 Frédéric 1995, p. 96.

181 Frédéric 1995, p. 100.

182 Ibid.

183 Huntingdon 1977, p. 64.

184 Ibid.

185 Ray 1999, p. 179.

186 De Saram 1966, p. 133.

187 Ray 1994, p. 179.

188 De Saram 1966, p. 11.

189 De Saram 1966, p. 11.

190 De Saram 1966, p. 69.

191 De Saram 1966, p. 91.

192 De Saram 1966, p. 90.

193 De Saram 1966, p. 52.

194 Keown 2004.

195 Ray, op.cit, p. 179.

196 De Saram 1966, p. 73.

197 De Saram 1966, p. 73.

198 Keown 2004, p. 339.

199 De Saram 1966, pp. 39–43.

200 De Saram 1966, p. 47.

201 De Saram 1966, p. 132.

202 De Saram 1966, pp. 141–142.

203 De Saram 1966, pp. 132–133. The nature of the sermon '*Adittha Pariyay Sutta*' by the Buddha remains unclear, a reference has yet to be found.

204 De Saram1966, pp. 67–69.

205 De Saram 1966, pp. 125,126.

206 De Saram 1966, pp. 120–123.

207 Greaves 2006, p. 23.

208 Bogle 2011, p. 67.

209 De Saram 1966, p. 83.

210 De Saram 1966, p. 14.

211 De Saram 1966, p. 62.

212 De Saram 1966, pp. 74,75.

213 De Saram 1966, pp. 71–73.

214 Ray 1994, p. 210.

215 Ray 1994, p. 179.

216 De Saram 1966, pp. 107,108.

217 Ray 1994, p. 205.

218 De Saram 1966, p. 50.

219 De Saram 1966, p. 45.

220 De Saram 1966, p. 25.

221 A reference to meaning has not been found.

222 A reference to meaning has not been found.

223 A reference to meaning has not been found.

224 A reference to meaning has not been found.

Part 6. Heavenly Texts: A Sampling

No references in this part.

Epilogue

225 King 1964, pp. 110, 111.

226 King 1964, pp. 108, 109.

Glossary and Alternate Spellings

227 Frédéric 1995, pp. 100–105.

228 Ray 1999, p. 179.

229 Huntingdon 1977, p. 64–71.

230 Donald K. Swearer. *Towards the Truth by Buddhadāsa.* Westminster Press, Philadelphia, 1971.

231 BhikkuBodhi. General Editor. *A Comprehensive Manual of the Abbidhamma.* Buddhist Publication Society. Kandy 1999.

232 Powers, John. *A Concise Encyclopedia of Buddhism.* Oneworld Publications. Oxford. 2000.

Other Sources

Keown, Damien. *Dictionary of Buddhism.* Oxford University Press. New York. 2004.

Nyanatiloka, Ven. *Buddhist Dictionary.* Frewin & Company. Colombo. 1972.

Schuhmacher, Stephan & Woerner, Gert, eds. *The Encyclopedia of Eastern Philosophy and Religion.* New York. Barnes and Noble. 1999.

Snelling, John. *The Buddhist Handbook.* Random House. London. 1998.

King Lithai. *Traibhumikatha: The story of the Three Planes of Existence.* Translated by the Thai National Team for Anthology of ASEAN Literatures. Published under the sponsorship of the ASEAN Committee on Culture and Information. Bangkok. 1985.

Frank E. & Mani B. Reynolds. *Three Worlds According To King Ruang a Thai Buddhist Cosmology.* Regents of the University of California. Berkeley. 1982.

BIBLIOGRAPHY

Alabaster, Henry. 1871. *The Wheel of Law: Buddhism*. London: Trübner.

Armstrong, Karen. 2001. *Buddha*. New York: Penguin.

Bechert, Heinz, and Richard Gombrich, eds. 1984. *The World of Buddhism*. London: Thames and Hudson.

Berry, Thomas. 1975. *Buddhism*. New York: Thomas Y. Crowell Co.

Blanchard, Wendell. 1958. *Thailand: Its People, Its Society, Its Culture*. New Haven: HRAF Press.

Blofeld, John. 1974. *The Tantric Mysticism of Tibet*. New York: Causeway Books.

Bodhi, Bhikku, ed. 1999. *A Comprehensive Manual of Abbidhamma*. Kandy: Buddhist Publication Society.

Bogle, James Emanuel. 2011. *Thai and Southeast Asian Painting: 18th through 20th Centuries*. Atglen, PA: Schiffer.

Brereton, Bonnie Pacala. 1995. *Thai Tellings of Phra Malai: Texts and Rituals Concerning a Popular Buddhist Saint*. Tempe: Arizona State University.

Caillat, Collette, and Ravi Kumar. 1981. *The Jain Cosmo*logy. New York: Crown.

Chaturachinda, Gwyneth, Sunanda Krishnamurty and Pauline W. Tabiang. 2004. *Dictionary of South and Southeast Asian Art*. Chiang Mai: Silkworm Books.

Coleman, J. A. 2007. *The Dictionary of Mythology*. New York: Arcturus.

Coomaraswamy, Ananda K. and Sister Nivedita. 1967. *Myths of the Hindus and Buddhists*. New York: Dover.

Cotterell, Arthur, and Rachel Storm. 1999. *The Ultimate Encyclopedia of Mythology*. London: Annes.

de Bary, William Theodore, ed. 1972. *The Buddhist Tradition in India, China and Japan*. New York: Vintage Books.

De Saram, C. 1966. *Pen Portraits of Ninety Three Eminent Disciples of the Buddha*. Taipei: Buddha Educational Foundation.

Evans-Wentz, W. Y. 2008. *Tibetan Book of the Dead*. New York: Metro Books.

Fischer-Schreiber, Ingrid, Franz-Karl Ehrhard, Kurt Friedrichs, and Michael S. Diener. 1999. *The Encyclopedia of Eastern Philosophy and Religion*. New York: Barnes and Noble Books.

Frédéric, Louis. 1995. *Buddhism*. Flammarion Iconographic Guides. Paris: Flammarion.

Garnier, Derick. 2004. *Ayutthaya: Venice of the East*. Bangkok: River Books.

Geographica: The complete illustrated Atlas of the world. 2004. Edited by Tom McKnight. New York: Barnes and Noble Books.

Greaves, Ron. 2006. *Key Words in Buddhism*. London: Continuum.

Herbert, Patricia. 1999. In *The Art of Burma: New Studies*, edited by Donald M. Stadtner, 89–102. Mumbai: Marg Publications.

Herbert, Patricia. 2002. "Burmese Cosmological Manuscripts." In *Burma: Art and Archaeology*, edited by Alexandra Green and T. Richard Blurton, 77–98. London: British Museum Press.

Grogerley, Daniel John. 1908. *Ceylon Buddhism*. London: Trubner.

Huntington, Harriet E. 1977. "The Sixteen Buddhist Arhats." *Arts of Asia* 7 (2): 62–71.

Huntington, John C., and Diana Bangdel. 2003. *The Circle of Bliss: Buddhist Meditational Art*. Chicago: Serindia.

Ions, Veronica. 1967. *Indian Mythology*. London: Paul Hamlyn.

Isaacs, Ralph, and T. Richard Blurton. 2000. Burma and the Art of Lacquer. River Books. Bangkok.

Jumsai, Sumet. 1989. Naga: Cultural Origins in Siam and the West Pacific. Singapore: Oxford University Press.

Kamsorn, Charun (Thai text), Roland D. Renard, et. al (English translation). 2006. A History of Kruba Sriwichai (The Buddhist Saint of Northern Thailand). Chiang Mai: Sutin Press.

Keown, Damien. 2004. Dictionary of Buddhism. New York: Oxford University Press.

Keyes, Charles F. 1989. *Thailand: Buddhist Kingdom as Modern Nation-State*. Bangkok: D. K. Book House.

King, Winston L. 1964. *A Thousand Lives Away: Buddhism in Contemporary Burma*. Cambridge, MA: Harvard University Press.

Kloetzli, Randy. 1983. *Buddhist Cosmology*. Delhi: Motilal Banarsidass.

Knappert, Jan. 1995. *Indian Mythology*. London: Diamond Books.

Lefferts, Leedom, and Sandra Cate. *Buddhist Storytelling in Thailand and Laos*. Asian Civilizations Museum. Singapore. 2012.

Lithai, King. 1985. *Traibumikatha: The Story of the Three Planes of Existence*. Translated by the Thai National Team for Anthology of ASEAN Literatures. Bangkok: ASEAN Committee on Culture and Information.

Lopez, Donald S., Jr. 2001. *The Story of Buddhism*. New York: Harper One.

Majupuria, Trilok, and Rohit Kumar. 2011. *Hindu, Buddhist and Tantric Gods, Goddesses, Ritual Objects and Religious Symbols*. Kathmandu: Scholar's Nest.

Matics, K. I. 1998. *Gestures of the Buddha*. Bangkok: Chulalongkorn University Press.

McDaniel, Justin Thomas. 2009. *Gathering Leaves and Lifting Words*. Chiang Mai: Silkworm Books.

McGill, Forrest, ed. 2005. *The Kingdom of Siam: The Art of Central Thailand, 1350–1800*. San Francisco: Asian Art Museum.

Nyanatiloka, Ven. 1972. *Buddhist Dictionary*. Colombo: Frewin.

Nyanatiloka, Ven. 1980. *Buddhist Dictionary: Manual of Buddhist Terms and Doctrines*. Kandy, Sri Lanka: Buddhist Publication Society.

Pal, Pratapaditya. 1990. "Arhats and Mahasiddhas in Himalayan Art." *Arts of Asia* 20 (1): 66–78.

Payutto, P. A. (Phraphom Khunaphon). 2007. *The Pali Canon: What a Buddhist Must Know*. Bangkok: Thammasapa and Bunluentham Printing.

Powers, John. 2000. *A Concise Encyclopedia of Buddhism*. Oxford: Oneworld.

Prakitnonthakan, Chatri. 2013. *The Philosophical Construct of Wat Arun*. Bangkok: The South East Insurance Company Limited.

Ray, Reginald A. 1999. *Buddhist Saints in India: A Study in Values and Orientations*. New York: Oxford University Press.

Reischauer, Edwin O., and John K. Fairbank. 1960. *East Asia: The Great Tradition*. Boston: Houghton Mifflin.

Reynolds, Frank E., and Mani B. Reynolds. 1982. *Three Worlds According to King Ruang: A Thai Buddhist Cosmology*. Berekley: Regents of the University of California.

Ridley, Michael. 1983. *Buddhism*. New Delhi: Heritage Publishers.

Rooney, Dawn. 2005. *Angkor*. Hong Kong: Airphoto.

Sadakata, Akira. 2009. *Buddhist Cosmology: Philosophy and Origins*. Tokyo: Kosei.

Saenkattiyarat, Suwat. 2007. *The Iconography of the Buddha*. Bangkok: SE-Education Public Company Ltd.

Snelling, John. 1998. *The Buddhist Handbook*. London: Random House.

Sri-Aroon, Khaisri. 2003. *Buddha Images in Thailand (Siam)*. Bangkok: Sujit Wongthes Publisher.

Tilakaratne, Asanga. 2012. *Theravada Buddhi*sm. Honolulu: University of Hawai'i.

Trai Phum Book: Ayutthaya Manuscripts–Thonburi Manuscripts. 1999. 3 vols. Bangkok: Amarin. In Thai. Published in honor of the King's 6th cycle (72nd year). [สมุดภาพไตรภูมิ ฉบับกรุงศรีอยุธยา–ฉบับกรุงธนชุรี. คณะกรรมการฝ่ายประมวลเอกสารและจดหมายเห็ตุ ในคณะอำนวยการจัดงานเฉลิมพระเกียรติพระบาทสมเด็จพระเจ้าอยู่หัว. จัดพิมพ์เนื่องในโอกาสพระราชพิธีมหามงคลเฉลิมพระชนมพรรษา **6** รอบ **5** ธันวาคม **2542]**

Vasubandhu. 1991. *Abhidharmakośabhāṣyam of Vasubandhu*. Vol. II. Translated into French by Louis de La Vallée Poussin. English Version by Leo M. Pruden. Berkeley: Asian Humanities Press.

USIS (United States Information Service). 1957. *The Life of The Buddha According to Thai Temple Painting*. Bangkok.

Wilkins, W. J. 1980. *Hindu Mythology Vedic and Puranic*. Delhi: Ruppa & Co. Delhi.

Wilkinson, Philip. 2010. *Religions*. New York: Metro Books.